REWARD

REWARD

Intermediate

Student's Book

Simon Greenall

MACMILLAN
HEINEMANN
English Language Teaching

Map of the book

Lesson	Grammar and functions	Vocabulary	Skills and sounds
1 *Could I ask you something?* Personal information; classroom language	Asking questions	Language learning and classroom language Words for giving personal information	**Reading:** reading and answering a questionnaire **Listening:** listening for main ideas **Sounds:** polite intonation in questions **Speaking:** talking about personal information
2 *Going places USA* Plans and arrangements	Present simple and present continuous	Journeys by train	**Listening:** listening with background noise; listening for main ideas **Writing:** completing a diary, writing a personal letter
3 *All dressed in red* Wedding customs in different countries	Describing a sequence of events (1): *before* and *after; during* and *for*	Words to describe weddings	**Reading:** reacting to a passage **Speaking:** talking about wedding customs **Writing:** writing about traditional weddings in your country
4 *Are you a couch potato?* Leisure activities	Adverbs (1): adverbs and adverbial phrases of frequency; talking about likes and dislikes; verb patterns (1): *to* or *-ing*	Adjectives to describe likes and dislikes Leisure activities	**Reading:** reading for main ideas **Listening:** listening for main ideas **Speaking:** talking about what you like doing in your spare time **Writing:** writing about other people's favourite leisure activities; linking words *and, but* and *because*
5 *Face the music* Different types of music	Adjectives (1): *-ed* and *-ing* endings; question tags	Types of music Adjectives to express how you feel about something	**Reading:** reading for main ideas; inferring **Sounds:** intonation in question tags to show agreement or to ask a real question **Speaking:** giving opinions about contemporary music **Writing:** writing a short report; linking words *on the whole, in my opinion, the trouble is, in fact*
Progress check lessons 1 – 5	Revision	Countries, nationalities and people Techniques for dealing with difficult words Starting a *Wordbank*	**Sounds:** stress change in words for countries and people; /ɪ/ and /i/; being aware of speaker's attitude **Speaking:** describing a photo **Writing:** joining sentences using *after* and *before* clauses; writing a paragraph about New Year's Eve
6 *How we met* Friendship	Past simple and past continuous	Nouns and adjectives to describe personal qualities	**Reading:** jigsaw reading for specific information **Listening:** listening for main ideas; listening for details **Writing:** writing a paragraph about a special friend
7 *The way things used to be* Childhood memories	*Used to* and *would* + infinitive	New words from a passage called *Investing in memories* Nouns and verbs which go together	**Reading:** reading for specific information **Sounds:** stress and intonation when correcting someone **Writing:** writing a paragraph about the way things used to be; linking words *when, after a while, eventually, now*
8 *Cold, lost, hungry and alone* Experiences in a foreign country	Describing a sequence of events (2): *when, as soon as, as, while, just as, until*	New words from a passage called *Cold, lost, hungry and alone*	**Listening:** listening for main ideas; listening for specific information **Reading:** reading for specific information **Writing:** writing paragraphs about experiences in a foreign country; linking words *when, as soon as, while, just as, until*
9 *Chocolate – like falling in love* Facts about chocolate	Non-defining relative clauses; *who, which, where*	Words related to food	**Reading:** reacting to a text; reading for main ideas **Speaking:** talking about well-known people, things and places **Writing:** writing about well-known people, things and places
10 *What did I do?* Social gaffes around the world	Verbs with two objects; complaining and apologising; making requests	New words from a passage called *What did I do?*	**Reading:** reading for main ideas **Listening:** listening for specific information **Sounds:** intonation in complaints and apologies **Speaking:** talking about misunderstandings
Progress check lessons 6 – 10	Revision	Adjectives and strong equivalents Adjective prefixes Homophones	**Sounds:** linking words in connected speech; word stress in sentences **Writing:** deleting and inserting words **Speaking:** talking about embarrassing incidents

Lesson	Grammar and functions	Vocabulary	Skills and sounds
11 *Do it now!* Organisational skills and routine activities	Present perfect (1): *already, yet, still*	Verbs for household actions	**Reading:** reading and answering a questionnaire **Listening:** listening for specific information **Speaking:** talking about domestic activities and personal organisation
12 *London calling* Radio news broadcasts	Present perfect continuous (1) for asking and saying how long	Words connected with news items; politics and military	**Listening:** listening for main ideas **Sounds:** word stress in sentences **Writing:** reconstituting a text **Speaking:** expressing opinions
13 *Fictional heroes never die* Famous characters from fiction	Present perfect simple (2) and present perfect continuous (2)	New words from a passage called *Fictional heroes never die*	**Reading:** reading for main ideas **Speaking:** talking about important events in your life **Writing:** writing a short biography
14 *Twin cities* A view of Prague	Making comparisons	Town features and facilities	**Listening:** inferring **Reading:** reading for specific information **Writing:** writing a description of a town, using linking words *the main reason why, another reason is that, both...and, neither...nor, while, whereas*
15 *I couldn't live without it* Personal possessions	Adjectives (2): order of adjectives	Possessions and objects around the home Compound nouns Nouns and adjectives which go together Techniques for dealing with unfamiliar words	**Reading:** reading for main ideas **Sounds:** stress in compound nouns **Speaking:** talking about personal possessions
Progress check lessons 11 – 15	Revision	Adjectives and participles Adjectives which go after the noun	**Sounds:** consonant clusters in end position; linking words in connected speech; stressed words in sentences **Speaking:** describing objects
16 *Do it in style* Style and fashion	Asking for and giving advice; *must* and *should*	Clothes, accessories, appearance Material and colours Techniques for dealing with unfamiliar words	**Reading:** reading for main ideas **Speaking:** talking about fashion **Writing:** writing advice for visitors to your country
17 *Rose Rose* A ghost story by Barry Pain	Making predictions: *may* and *might*; *going to* and *will*	New words from a story called *Rose Rose*	**Listening:** listening for main ideas; listening for specific information **Speaking:** predicting **Writing:** writing the ending of a story from a different point of view; linking words *although, in spite of + -ing*
18 *What do you do for a living?* Jobs and conditions of work	Drawing conclusions; *must, can't, might, could*; describing impressions	Jobs Adjectives to describe personal and professional qualities	**Speaking:** talking about your job or your ideal job **Sounds:** stressed syllables in words; intonation when disagreeing **Listening:** listening for specific information **Reading:** understanding text organisation
19 *Guided tours* Pilgrimage to Santiago de Compostela	Talking about obligation, permission and prohibition	Religion and tourism	**Reading:** reading for main ideas; inferring **Sounds:** strong and weak forms of *can, have to* and *must* **Writing:** a description of rules in your country
20 *How unfair can you get?* Emotional reactions	Talking about ability and possibility: *can, could, be able to*	Compound adjectives	**Speaking:** talking about personal qualities **Listening:** predicting; listening for main ideas; listening for detail; listening for specific information **Writing:** re-telling the story from notes and from the listening activity
Progress check lessons 16 – 20	Revision	Multi-part verbs: phrasal verbs and prepositional verbs with objects	**Sounds:** different ways of pronouncing the letter *a*; linking words in connected speech; consonant clusters in initial position **Speaking:** talking about unlikely statements **Writing:** writing statements which must be, might be and can't be true

Lesson	Grammar and functions	Vocabulary	Skills and sounds
21 *Cinema classics* Films	Adverbs (2): formation; giving opinions; emphasising	Types of films Adjectives to describe opinions	**Listening:** listening for specific information; listening for main ideas **Sounds:** strong intonation for emphasising an opinion **Writing:** inserting words; writing the plot of a classic film **Speaking:** talking about classic films
22 *Wild and beautiful* Animal conservation	Adverbs (3): position of adverbs and adverbial phrases	Animals	**Reading:** reading for main ideas; evaluating a text **Speaking:** talking about environmental changes in the world **Writing:** writing about something which has changed with linking words *ago, then, today, in ...years' time*
23 *Valentine* A poem	Reported speech (1): statements	Love symbols Techniques for dealing with unfamiliar words Adjectives to describe personal characteristics	**Reading:** reading for specific information **Listening:** inferring; listening for main ideas **Sounds:** reciting a poem **Speaking:** talking about romantic things, people and places **Writing:** writing a poem
24 *Medium wave* The media	Reported speech (2): questions	Newspapers, television and radio	**Listening:** listening for specific information **Writing:** writing a report of an interview; linking words *the reason for this, because, another reason for this, besides* **Speaking:** talking about the media
25 *A cup of tea* A short story by Katherine Mansfield	Reported speech (3): reporting verbs	New words from a story called *A cup of tea*	**Speaking:** predicting **Reading:** predicting; reading for main ideas **Writing:** rewriting the story from another point of view
Progress check lessons 21 – 25	Revision	Multi-part verbs: phrasal verbs without objects and phrasal prepositional verbs	**Sounds:** different ways of pronouncing the letter *e*; silent consonants; stress when correcting someone **Speaking:** talking about yourself; responding to others
26 *Eat your heart out... in the USA* American regional cooking	Giving instructions and special advice	Food Ways of preparing food, ways of cooking, kitchen equipment	**Speaking:** talking about typical food from your country or region **Reading:** reading for specific information; reacting to a text **Listening:** listening for specific information; understanding text organisation
27 *Home thoughts from abroad* Holiday postcards	Defining relative clauses	Holidays	**Speaking:** talking about holiday postcards **Reading:** reading for main ideas; linking ideas **Sounds:** pauses in relative clauses **Listening:** listening for main ideas **Writing:** writing postcards
28 *Local produce* Things grown and made in Britain	The passive	Manufactured and natural products Words which go together Techniques for dealing with unfamiliar words	**Reading:** reading for main ideas **Speaking:** talking about typical food, drink or products of your town, region or country **Writing:** writing about typical food, drink or products
29 *Just what we're looking for!* Looking for somewhere to live	Verb patterns (2): *need + -ing* and passive infinitive; causative constructions with *have* and *get;* reflexive pronouns	Adjectives to describe things which go wrong with a house Furniture and fittings Jobs Verbs for household maintenance	**Reading:** reading for main ideas **Speaking:** discussing if you are houseproud; talking about homes
30 *Sporting chance* New rules for old sports	Verb patterns (3): *make* and *let;* infinitive constructions after adjectives	Sports and sports equipment Words which go together	**Reading:** inferring **Listening:** listening for main ideas **Sounds:** stressed words in sentences **Speaking:** talking about sport
Progress check lessons 26 – 30	Revision	Verbs and nouns which go together Words which are sometimes confused	**Sounds:** different ways of pronouncing the letter *i* **Reading:** understanding text organisation **Writing:** inserting words and sentences

Lesson	Grammar and functions	Vocabulary	Skills and sounds
31 *I never leave home without it* Useful objects and equipment	Zero conditional: *in case*	Personal possessions	**Listening:** listening for main ideas **Sounds:** syllable stress **Reading:** reading for main ideas **Speaking:** talking about the things you always take with you
32 *Politely but firmly* Letters Complaining about a new car	Describing a sequence of events (3): *as soon as, when* and *after* for future events	Parts of a car	**Speaking:** talking about making complaints **Reading:** reading for main ideas **Listening:** listening for main ideas **Writing:** writing a letter of complaint
33 *Superhints* Practical household advice	Verb patterns (4): infinitive of purpose; *by + -ing*; giving advice; *if* clauses	Household objects and actions	**Reading:** reading for specific information **Sounds:** different pronunciation of vowels and diphthongs; weak forms in connected speech **Listening:** listening for specific information **Writing:** writing household advice
34 *The green tourist* The effects of tourism on the environment	First conditional	New words from a passage called *Are you a green tourist?* Words which go together	**Reading:** reading for main ideas, inferring **Sounds:** /ɪ/; stress and intonation in first conditional sentences **Listening:** listening for specific information **Speaking:** talking about the effects of tourism
35 *Lost in the Pacific* The story of Amelia Earhart	Past perfect simple and continuous	New words from a passage about Amelia Earhart	**Reading:** reading for main ideas; understanding text organisation **Writing:** writing a paragraph about an unsolved mystery
Progress check lessons 31 – 35	Revision	English words borrowed from other languages Words which go in pairs Antonyms	**Sounds:** different ways of pronouncing the letter *o*; syllable stress **Speaking:** talking about true and false information about Britain **Listening:** listening for main ideas
36 *What's your advice?* Letters to a problem page	Second conditional; giving advice	Relationships	**Reading:** reading for main ideas **Sounds:** linking words in connected speech **Listening:** listening for main ideas **Writing:** writing a letter of advice
37 *You should have been here last week!* A visit to Hong Kong	Past modal verbs (1): *should have*	Tourism in Hong Kong Techniques for dealing with unfamiliar words	**Reading:** reading for main ideas; inferring **Listening:** listening for main ideas; listening for detail **Speaking:** talking about something you did wrong and which you regret
38 *Now you see me, now you don't* Strange and supernatural incidents	Past modals (2): *may have, might have, could have, must have, can't have*	New words from three stories about strange incidents	**Reading:** understanding text organisation; inserting phrases **Speaking:** telling stories; speculating
39 *Making the grade* Education in Britain and the USA	Expressing wishes and regrets	Education Subjects that are studied at school or college	**Listening:** listening for specific information **Sounds:** /d/; intonation in sentences with *if only* **Speaking:** talking about important events in your educational career **Writing:** writing a description of the education system in your country; linking words *like* and *unlike*
40 *The man who was everywhere* A thriller by Edward D Hoch	Third conditional	New words from a story called *The man who was everywhere*	**Reading:** understanding text organisation; understanding a writer's style **Listening:** listening for main ideas **Speaking:** predicting; discussing what you would have done **Writing:** rewriting the story from a different point of view
Progress check lessons 36 – 40	Revision	Colloquial language Idiomatic expressions Techniques for dealing with unfamiliar words	**Sounds:** different ways of pronouncing the letter *u*; silent letters; linking words in connected speech **Reading:** understanding text organisation **Speaking:** reacting to a story

1 | *Could I ask you something?*

Asking questions

VOCABULARY AND READING

1 Write down six pieces of information about yourself using the words and phrases in the box below.

> married be born birthday children old divorced
> nationality drink father favourite first name wife
> surname food brother flat get up single go to bed
> go to work have dinner have lunch sister husband
> address house live mother work spare time sport

My surname is Borges. My favourite food is pizza.

2 Work in pairs and show each other what you wrote in 1. Tell the others in your class about your partner.

3 The words in the box below are often used for instructions in the classroom and in *Reward* Intermediate. Underline the verbs.

> activity answer check complete conversation do
> explain hear look up listen make meaning
> mistake spell match notes number paragraphs
> phrase chart pronounce work aloud question
> read repeat circle say exercise in pairs sentence
> ask statement tick underline word passage write

Which words in the box do the verbs go with?

4 Complete these instructions from *Reward* Intermediate with words from the box in activity 3.

1 Read the ___ and match the names with the ___ . (READING)
2 Here are some new ___s from the story. ___ you understand what they mean. (VOCABULARY)
3 ___ these sentences with *who, where* or *which*. (GRAMMAR)
4 Listen and ___ anything in the ___ that is different from what you ___ . (LISTENING)
5 ___ and say these words ___ . (SOUNDS)
6 Write ___ in ___ to these questions. (WRITING)
7 Work ___ . Ask and ___ what you like doing. (SPEAKING)

5 Read the questionnaire about asking questions in the classroom and decide what you say. (Think about grammar as well as what is polite and friendly.)

Could I ask you something?

1 You want to use your friend's pen. What do you say?
a 'Give me your pen.' ☐
b 'I'd like your pen.' ☐
c 'Can I use your pen, please?' ☐

2 You don't know what an English word (for example, *chart*) means. What do you say?
a 'What means *chart*?' ☐
b 'What does *chart* mean?' ☐
c 'Can I go now?' ☐

3 There are over twenty words in a lesson which are new to you. What do you say to your teacher?
a 'Could you explain what they mean?' ☐
b 'Which words should I write down?' ☐
c 'When does this lesson finish?' ☐

4 Your teacher says something to you very quickly. What do you say?
a 'Could you repeat that, please?' ☐
b 'What?' ☐
c 'Slow down!' ☐

5 You don't know how to say something (for example, *salut*) in English. What do you say?
a 'How you say *salut* in English?' ☐
b 'How do you say *salut* in English?' ☐
c '*Salut* in English.' ☐

6 You don't know how to spell an English word (for example, *sentence*). What do you say?
a 'How do you spell *sentence*?' ☐
b 'Could you write *sentence*?' ☐
c 'How you write *sentence*, please?' ☐

7 Your teacher asks you to read a difficult passage. What do you say?
a 'Could you pass me the dictionary?' ☐
b 'I wonder if you could tell me what these words mean?' ☐
c 'Would you mind explaining the general sense of the passage?' ☐

8 Your teacher plays you a listening text which is very difficult. What do you say?
a 'It's too difficult.' ☐
b 'Could you write up some important words and then play it again, please?' ☐
c 'Help!' ☐

LISTENING

1  Listen to Gail, who is an English teacher, and Tony, who is French, answering the questionnaire in *Vocabulary and reading*. What are her answers? Do you agree?

2 Work in pairs. Why were the other answers wrong, or less polite?

Listen again and check.

GRAMMAR

Asking questions

You form questions in two ways:
– without a question word and with an auxiliary verb, eg *be*, *do* or *have*. The word order is auxiliary + subject + verb.
Are you married?   Was she born in London?
Do you have any brothers?
Did you get up late this morning?

– with a question word, eg *who*, *what*, *where*, *how*, *why*, *when* and an auxiliary verb. The word order is question word + auxiliary + subject + verb.
What nationality are you?   Where were you born?
How do you say 'salut'?   What do these words mean?

You can form more indirect, polite questions with one of the following question phrases.
Can I *use your pen, please?*
Could you *repeat that, please?*
Would you mind *explaining the general sense, please?*
Could you *tell me how you say 'salut'?*
I wonder if you could *tell me what these words mean?*
In the last two question phrases, the word order is question phrase + subject + verb.

---

1 Put these words in the correct order and make sentences.

1 were you where born
2 nationality what you are
3 tell your could what is you me name
4 repeat could you that please
5 would mind you more slowly speaking please
6 you wonder I if could means me what tell 'spare-time'

1   Where were you born?

2 Think of four or five names, dates and places which are important in your life and write them down. Use some of the words for giving personal information in the vocabulary box to help you.

Thomas, 1975, Lyon,

3 Now work in pairs. Show your partner the information you wrote in 2. Ask and answer direct questions about each other. Use questions without question words.

Were you born in 1975?   Yes, I was.
Do you live in Lyon?   No, I don't.
Is your son's name Thomas?   Yes, it is.

4 Work in pairs and check your answers to the questionnaire.

SOUNDS

Listen to the questions in *Grammar* activity 1. Put a tick (✓) if the speaker sounds polite and friendly.

Now say the questions aloud. Try to sound polite and friendly.

SPEAKING

1 Work in pairs. Think of five things you would like to know about your partner. Are there any questions which it isn't polite to ask people you don't know well in your country?

2 Ask and answer questions about each other. Use polite question forms.

2 | *Going places USA*

Present simple and present continuous

VOCABULARY AND LISTENING

1 🔲 You are going to hear three interviews with people in Grand Central Station in New York. First, listen to the background noise and decide where the interviews take place.

– at the entrance   – at the ticket office
– on the platform   – in a waiting room

2 Work in pairs and check your answers to 1. Describe the situations with some of these words.

> luggage announcement one-way
> arrival booking office cab cab rank
> cancel connection traffic crowds
> delay departures fare information desk
> leave catch passenger reservation
> baggage round trip suitcase ticket
> coach timetable car horn

3 🔲 Listen to the interviews with Fran Ramirez, Henry North, and Joanie and Stephen Goldberg. As you listen, put the number of the interview in which you hear these phrases.

leave for home	3
spend Christmas with grandchildren	3
wait for a train	1
have a cup of coffee	1
take the train	3
walk to work	1
have a rehearsal	2
take a cab	2
work for an advertising agency	1
play the cello	2
teach music	2
work on Wall Street	3

4 🔲 Now listen again and check your answers to 3.

GRAMMAR

> ### Present simple
> **You use the present simple to talk about:**
> **– a general truth, such as a fact.**
> *Half a million people pass through the station daily.*
> **– something that stays the same for a long time, such as a state.**
> *He lives in Queens.*
> **– something that is regular, such as routines, customs and habits.**
> *They usually spend Christmas in New York.*
>
> ### Present continuous
> **You use the present continuous to talk about:**
> **– an action which is happening at the moment.**
> *She's having coffee.*
> **– an action or state which is temporary.**
> *He's working at the Met.*
> **– a definite arrangement in the future.**
> *They're spending Christmas in Chicago.*
> **You often use the present continuous for arrangements with verbs like *arrive, come, go, leave, meet, see, visit.***
> **Remember there are some verbs you don't usually use in the present continuous. Here are some of them: *believe, hate, hear, know, like, love, smell, sound, taste, understand, want.***

1 Answer the questions.

Fran Ramirez
1 What does she do?
2 What's she doing at the moment?
3 How does she get to work?
4 Is she taking the train or walking at the moment?

Henry North
1 What's his usual job?
2 What's he doing at the moment?
3 How does he usually get to the Met?
4 How is he getting to work today?

2 Write full answers to these questions about Joanie and Stephen's arrangements. Use the present continuous.

1 Where are they spending Christmas?
2 What train are they taking?
3 What are they going to do?
4 What time are they arriving in Chicago?

3 Write sentences about yourself or your town, showing each use of the present simple and continuous described in the box.

A million people live in my home town. (a general truth)
I go to work every day. (a routine)

WRITING

1 Read Joanie and Stephen's letter to their son and daughter-in-law and make a list of their arrangements.

> 1041, Penn Street
> Staten Island
> New York, New York
> 16 November
>
> Dear Maisie and Tom,
> Well, it's all arranged! We're spending Christmas with you and then we're going to Toronto to see Gary and Holly on 27 December. We're staying with them for three days, then we're returning home on 31 December in time for New Year at home.
> We're coming to Chicago by train because we hate flying, as you know.
> We hope you can meet us in Chicago.
> Love
> Mom and Pop

Christmas – Maisie and Tom

2 Look at the personal letter in 1 and answer the questions.

1 Whose address is at the top?
2 How do you start a personal letter?
3 Where do you put the date?
4 How do you finish it?

Which of the following expressions are other possible ways of finishing a personal letter?

Best wishes   Kind regards
Yours sincerely   Yours faithfully

Do you lay out letters in the same way in your language?

3 Make a list of your arrangements for the next week.

Monday: lunch with Helena
Tuesday: meeting

4 Write someone a letter about your arrangements for next week. Use the letter in 1 as a model.

3 | *All dressed in red*

Describing a sequence of events (1); *before* and *after*; *during* and *for*

VOCABULARY AND SPEAKING

1 Work in pairs. The theme of this lesson is weddings in different countries. Which of the following words can you use to describe the preparations for a wedding, the ceremony and the celebration in your country?

best man bride bridesmaid cake cash ceremony Christian church couple engagement gold groom henna Hindu honeymoon horoscope jewel marriage matchmaker Moslem paste priest promise propose reception registry office ring sari Taoist town hall veil witness

2 Talk about wedding customs and traditional ceremonies in your country.

The groom arrives at the church with the best man. Then the bride arrives with her father.

READING

1 Read the passage and find out why it's called *All dressed in red*.

2 Does the passage say anything about the following? If so, what does it say?

– preparations
– presents
– dress
– reception
– wedding ceremony

Before getting married, the bride at a Hindu wedding ceremony does everything to make sure her wedding day is a lucky one. A holy man studies the horoscopes of the bride and groom to choose the right day for the wedding, so that the marriage will be long and happy. Will the marriage be a happy one if the wedding takes place on Wednesday, or would it be better to wait until the full moon on Friday? After checking all the signs very carefully, he chooses the wedding day. There are the weeks of preparation and excitement that are common to all cultures, then finally, on the wedding day itself, before helping her dress, the bride's sisters and female friends paint her hands and feet with henna. When she's ready, she puts on a red sari, the colour which will bring her good luck, for the marriage. During the ceremony, the groom's relatives place a small mark of red paste on her forehead to show she is a married woman. After decorating the bride with jewels, they cover both her face and that of the groom with a veil. And then she is married. During the reception, the guests enjoy a feast of food and drink, while the bride and groom sit together and share their meal.

First printed in British Airways *High Life*

All dressed in red

3 Work in pairs.

Student A: Turn to Communication activity 1 on page 98.
Student B: Turn to Communication activity 10 on page 100.

4 Work in pairs and tell each other about weddings in Moslem countries and in China. What do the passages say about the things in 2? What do you find most interesting or different from weddings in your country?

FUNCTIONS AND GRAMMAR

> **Describing a sequence of events**
>
> You use *before* or *after* to link two actions. You use *before/after* + subject + verb when the subject is the same or different in the two actions.
> *Before they help her dress, they paint her hands and feet.*
> **(They paint her hands and feet. Then they help her dress.)**
> *After he has made sure that the signs are good, the two families ask the gods for their help.*
> **(He makes sure the signs are good. Then the two families ask the gods for help.)**
> You can also use *before/after* + *-ing* when the subject is the same in the two actions.
> *Before helping her dress, they paint her hands and feet.*
> **(They paint her hands and feet. Then they help her dress.)**
> *After arriving at the wedding reception, they sit on a small stage.*
> **(They arrive at the wedding reception. Then they sit on a small stage.)**
> You can put the following phrases in front of *before* and *after*: *a week, a day, a year, a few days, shortly, just, less than a week, for weeks.*
> *For weeks before choosing the day, the holy man has been studying the horoscopes.*
>
> *During* and *for*
> You use *during* to say when something happens.
> *During this part of the wedding, they receive and open the wedding presents.*
> You use *for* to say how long something takes.
> *The celebration lasts for several hours.*

1 Look back at *All dressed in red* and underline the *before* and *after* clauses.

2 Read *All dressed in red* and the two passages in *Reading* activity 3 again and join these sentences. Use *before* or *after* + subject + verb at the beginning of the first sentence.

1 The Hindu holy man checks the horoscopes.
  The Hindu holy man chooses the wedding day.
2 Friends and relatives paint the Hindu bride's hands and feet.
  The Hindu bride puts on her sari.
3 The groom's relatives cover the bride and groom with a veil.
  The groom's relatives decorate the bride with jewels.
4 The Moslem couple celebrate their wedding together.
  The groom attends a religious ceremony.
5 They arrive at the reception.
  They receive the wedding presents.
6 The matchmaker takes a present to the Chinese bride.
  The groom can propose to his bride.
7 They agree to the marriage.
  They check the couple's horoscopes.
8 They ask the gods for their help.
  The matchmaker makes sure the signs are good.

1 After the Hindu holy man checks the horoscopes, he chooses the wedding day.

3 Rewrite any sentences in 2 which can take *before/after* + *-ing*.

4 Complete these sentences with *during* or *for*.

1 ___ the ceremony, they exchange rings.
2 The reception may last ___ two days.
3 Preparations go on ___ several months.
4 ___ the reception the couple receives presents.
5 She wears the red mark ___ the rest of her married life.
6 The holy man studies the horoscopes ___ this period.

SPEAKING AND WRITING

1 Work in groups of two or three. Are there any unusual weddings or alternatives to traditional weddings in your country? Is marriage in your country as popular as it was fifty years ago? What reasons are there to get married?

2 Write down the different stages of preparation, ceremony and celebration in traditional or alternative wedding ceremonies in your country.

In Britain, the groom arrives at the church with the best man. They sit on the right-hand side. The bride arrives with her father.

3 Join your notes with any other ideas from 1, and write a paragraph about traditional or alternative weddings in your country.

After arriving at the church, the groom and the best man sit on the right-hand side. The bride arrives with her father...

4 *Are you a couch potato?*

Adverbs (1): adverbs and adverbial phrases of frequency; talking about likes and dislikes; verb patterns (1): *to* or *-ing*

VOCABULARY AND READING

1 Look at these words to express likes and dislikes. Put them in order from positive to negative.

> all right awful boring brilliant dreadful dull exciting fun great nice
> OK relaxing superb terrible terrific wonderful

2 Look at the leisure activities in the box below. What's your opinion of them? Use the words in the box in 1.

> football tennis cricket do-it-yourself (DIY) fishing gardening entertaining
> shopping going to nightclubs watching television reading painting
> bird-watching train spotting playing cards swimming running walking

I think football is boring.

Which other leisure activities do you like and dislike?

3 *Are you a couch potato?* is about leisure activities. Read it and choose the best definition for a *couch potato*.

1 Someone who enjoys energetic sports and active hobbies.
2 Someone who takes little or no exercise, and who spends their free time doing very little.
3 Someone who doesn't like doing sports but is active in other ways.
4 Someone who likes indoor gardening.

4 Work in pairs. Decide how energetic the activities in the box in 2 are. Give them a score from 1 (= lazy), to 10 (= energetic).

LISTENING

1 [cassette icon] Listen to nine people talking about what they like doing. Put the number of the speaker by the activity in the box in *Vocabulary and reading* 2 which they like doing.

2 Work in pairs and decide how old each person might be. Which person or people would you describe as a *couch potato*? What else did they say about their leisure activities?

[cassette icon] Listen again and check.

Are you a couch potato?

Centuries ago, people didn't have much free time, because everybody was working too hard. In Britain in the nineteenth century, people had more spare time, but because the Victorians hated relaxing and doing nothing, they invented football, rugby and cricket. People took up more gentle activities too, like gardening, bird-watching and train spotting, and it was even possible simply to watch a sport and give the impression that you were actually doing something. Gradually, leisure activities have become less and less demanding, and most people have a variety of more or less energetic interests and hobbies. But now there is a new type of person who thinks that lying on the sofa watching television on Sunday afternoon or reading the newspaper from cover to cover is the most exciting activity they can manage. This is the twentieth-century couch potato. For them, every activity is too much trouble, and laziness is an art form! So how do you spend your free time?
Are you a couch potato?

FUNCTIONS AND GRAMMAR

Adverbs and adverbial phrases of frequency
Adverbs of frequency usually go before a full verb, but after _be_ or an auxiliary verb.
I **always** spend the weekend doing housework.
She **sometimes** likes playing cards.
I am **often** out in the evening.
I've **never** enjoyed football.
I **hardly ever** go to the cinema.

Here are some common adverbial phrases of frequency. They usually go at the end of a clause.
every day, week, month, year, two days, other day, now and then
once/twice/three times a day, a week, a month, a year
most days, most mornings, once in a while

Talking about likes and dislikes
You can put an _-ing_ form verb or a noun after the following expressions.
I **adore** shopping.   _I **love** entertaining._
I **enjoy** watching television.
I **hate** running.   _I **can't stand** staying at home._
I **detest** collecting the leaves.
I **don't mind** spending Saturday with the children.

Verb patterns (1): _to_ or _-ing_
Like and _love_ + _-ing_ means _enjoy doing something._
I like going shopping. = I enjoy it.

Like and _love_ + _to_ + infinitive suggests that you choose to do something because it's a good idea.
You may or may not enjoy it as well.
I like to go shopping on Mondays. = Mondays is the best time for me to go shopping.

Remember that you can use _would love/like to_ + infinitive to talk about ambitions, hopes or preferences.
I'd love to learn to ski. I'd like to have lessons this winter.

1 Write sentences saying how you feel about these activities.

– shopping   – swimming
– bird-watching   – train spotting

2 Choose the best verb pattern. If two answers are possible, explain the difference in meaning.

1 I like _to go/going_ to the dentist every six months.
2 She likes _to get/getting_ home before it's dark.
3 She likes _to visit/visiting_ her parents.
4 He likes _to do/doing_ the washing on Mondays.

3 Write sentences saying how often you do the following activities. Make sure you put the adverb or adverbial phrase in the right position. ·

1 have a holiday   3 listen to the radio
2 read a newspaper   4 go to the cinema

SPEAKING AND WRITING

1 Work in groups of three or four and find out if there are any couch potatoes in your class.

> **1** Each person make a list of your five favourite free time activities.
>
> **2** Give each activity a score from 1 (lazy) to 10 (energetic).
>
> **3** Add up the scores for your activities. Is the person with the lowest score a couch potato?

2 Go round the class asking and talking to people in other groups about your favourite and least favourite leisure activities. Find out if there are any other couch potatoes in your class.

3 Write a paragraph about what three or four people in your class enjoy doing. Use linking words _and_, _but_ and _because_.

– Say what they like doing.
Frederico likes going to football and tennis matches...

– Say why they like doing it.
... because he thinks they're very exciting.

– Say if they dislike anything.
... but he can't stand the crowds.

– Say how often they enjoy doing it.
He goes five or six times a year.

5 *Face the music*

**Adjectives (1): *-ed* and *-ing* endings;
question tags**

VOCABULARY

1 Work in pairs. Do you know the
types of music in the box below?

> rap rock heavy metal reggae
> country and western folk
> classical opera funk jazz
> techno blues pop soul salsa
> middle-of-the-road

Listen to six pieces of
music. Put the number of each
piece of music by its type in the
box above.

Are there any types of music
which you can add to the list?
What type of music is most
popular in your country at the
moment?

2 Work in pairs. Talk about your
favourite and least favourite
types of music, pieces of music
and performers. You can use the
words below and ways of talking
about likes and dislikes from
Lesson 4.

> amusing boring depressing
> exciting great interesting
> irritating moving passionate
> relaxing sad silly thrilling

*I think opera is incredibly
boring, but I love rap.*

3 Which adjectives in the box come
from a verb?

The Beatles, Adriano Celentano, Julio Iglesias, Johnny Halliday, Elvis Presley – all
names from the days when everyone knew what music they liked, and everyone
liked the same music. But what about music of today? Do we understand it or do
we just love the noise it makes?

'No one is interested in the lyrics today, are they? It's the music that's important.'
Eva, Budapest

'The golden age of rock music was the sixties and seventies, wasn't it? Since then, it's so
frustrating that there are no great groups today, which are likely to be around in twenty
years' time.' *Gunther, Hamburg*

'I'm astonished how popular opera is now. I think it's due to singers like Pavarotti and
Domingo.' *Sinead, Cork*

'Julio Iglesias is still the greatest, isn't he? He's so relaxing. I don't understand what he's
saying, but it doesn't matter, does it?' *Sally, Brighton*

'I'm bored with so much British and American rock. Why can't we hear more music from
other countries and cultures?' *Jeanne, Paris*

'The trouble is, there's too much choice these days. No one can possibly keep
up to date with all the new bands, can they?' *Sandra, Buenos Aires*

'You don't have to know much about music to be a pop musician these days, do you? There
aren't many bands that can play their own instruments. It's all done in the studio, isn't it?'
Ingrid, Goteborg

READING

1 Read *Face the music* and decide if you agree with the speakers.

2 Work in pairs and decide if these statements about the passage are true or false, or if the passage doesn't say.

1 Eva thinks the music is more important than the words.
2 Gunther thinks today's music is not very good.
3 Sinead likes opera.
4 Sally doesn't listen to the words of Julio Iglesias' songs.
5 Jeanne likes a broad variety of music.
6 Sandra doesn't like recent music.
7 Ingrid thinks many rock bands can't play their own music.

GRAMMAR

Adjectives ending in -ed and -ing

Many adjectives formed from past participles describe a feeling or a state.
I'm bored. = I feel there is nothing that interests me at the moment.
Many adjectives formed from present participles describe the person, thing or topic which produces the feeling.
I'm boring. = I'm a very uninteresting person.
Other adjectives like this are: shocked – shocking, worried – worrying, surprised – surprising, embarrassed – embarrassing, annoyed – annoying.

Question tags

Question tags turn a statement into a question. If the statement is affirmative, you use a negative tag. If the statement is negative you use an affirmative tag.
You like jazz, don't you? You don't like opera, do you?
Look at the replies to the questions.
Julio Iglesias is the greatest, isn't he? - Yes, he is.
= I agree, he's the greatest.
It doesn't matter, does it? - No, it doesn't.
= I agree, it doesn't matter.
Question tags can be used to ask a real question. In this case the intonation rises on the tag. Often question tags can be used to show friendliness or make conversation. In this case, you are not asking a real question, but showing you expect agreement, and the intonation falls on the tag.

1 Complete the sentences with adjectives ending in *-ed* or *-ing* formed from the verb in brackets.

1 Classical music is ___ . (bore)
2 He gets very ___ when he listens to heavy metal. (excite)
3 It's ___ to listen to his singing. (embarrass)
4 I feel ___ when I listen to jazz. (relax)
5 I find opera quite ___ . (thrill)
6 What a ___ song! (depress)

2 Complete the sentences with question tags.

1 You're going to the concert, ___ ?
2 This isn't my guitar, ___ ?
3 She plays in a band, ___ ?
4 He doesn't sing very well, ___ ?
5 They've got lots of money, ___ ?
6 You like listening to country and western, ___ ?

SOUNDS

1 🔊 You will hear the sentences in *Grammar* activity 2. Write *R* if you think the intonation is rising and *F* if you think the intonation is falling.

2 Say the sentences aloud, first with a rising intonation to ask real questions, then with a falling intonation showing you expect agreement.

SPEAKING AND WRITING

1 Find out what other people in the class think about the speakers' opinions in *Face the music*. Do you agree or disagree with them? Use question tags if possible.

2 Write a short report about people's opinions in 1. Use the linking words in **bold**.

Give a general opinion.
On the whole, *we think today's pop music is great.*

Give some more opinions.
*But **in my opinion**, no one listens to the lyrics.*

Describe any criticisms.
The trouble is, *you can't hear the words.*

Emphasise any points.
*...and **in fact**, some performers can't even play.*

Progress check 1–5

VOCABULARY

1 You add *-s* to most *adjectives of nationality* to form the word for the people from a particular country.

country	nationality	people
Argentina	Argentinian	the Argentinians
Brazil	Brazilian	the Brazilians

With adjectives ending in *-sh, -ch, -s* and *-ese*, you don't add *-s*.

country	nationality	people
China	Chinese	the Chinese
Denmark	Danish	the Danish
England	English	the English
France	French	the French

There are a few irregular forms. The people from Holland are called the Dutch and the people from Scotland the Scots (Scottish is the adjective and Scotch is only the drink).

Complete the table below.

country	nationality	people
Germany		
Hungary		
India		
Japan		

Continue the chart with countries, people and nationalities for the rest of the alphabet (if possible).

2 Often when you are reading English you will come across words that you may not understand. If you do not understand a word, there is a series of techniques you can use to guess its meaning.

- Decide what part of speech the word is, and then look carefully at the word or words it goes with. You may know these and may be able to guess the difficult word.

- Decide if the word is important for the general meaning of the passage. For example, you may not have to guess the exact meaning of words which are in lists, in brackets or preceded by *for example*.

- Look at the rest of the sentence or paragraph. The meaning may become clear from the context.

- Guess the main idea of the word without finding its exact meaning.

- Read on and confirm or revise your guess.

3 The vocabulary you will learn most easily are the words and phrases which are most useful to you. Look at the vocabulary boxes in Lessons 1–5 and decide which words are most useful to you.

When you are ready, you can start a *Wordbank* in your Practice Book, where there is space for you to write down new words in categories to help you learn them effectively.

GRAMMAR

1 Here are some answers to questions about problems in the classroom. Write the questions.

1 I-T-A-L-I-A-N.
2 The English for *heureux* is *happy*.
3 /əmerɪkn/
4 *Impatient* means *not patient*.
5 It means *say again*.

2 Write sentences saying how often you do the following activities.

1 look through your vocabulary notes
2 revise your grammar
3 read something in English
4 listen to spoken English
5 speak English outside the classroom
6 write something in English

3 Choose the correct verb form.

We *go/are going* to Italy next week. We *take/are taking* the plane. Usually we *have/are having* two weeks' holiday, but this year we *have/are having* four. We *don't take/aren't taking* the car because the journey *takes/is taking* too long. We *stay/are staying* at a friend's house in Varese. Usually he *works/is working* in Milan, but at the moment he *spends/is spending* the summer in New York.

4 Rewrite these sentences with *after + -ing*.

1 I get dressed. Then I have breakfast.
2 I have a cup of coffee. Then I start work.
3 I say hello to my friends. Then I go and see my boss.
4 I open my mail. Then I make some phone calls.
5 I check my appointments for the next day. I leave the office at 5pm.
6 I have a drink with friends. Then I go home.

5 Rewrite the sentences in 4 with *before + -ing*.

6 Complete these sentences with the *-ed* or *-ing* adjective formed from the verb in brackets.

1 I found the film very ___ . (disappoint)
2 It's a ___ book. (bore)
3 I get ___ by rude people. (annoy)
4 I find motorway driving very ___ . (frighten)
5 The children are ___ about the holidays. (excite)
6 He was ___ to hear your news. (delight)

7 Complete these sentences with question tags.

1 Marie is French, ___ ?
2 You speak Italian, ___ ?
3 She doesn't drive, ___ ?
4 You've got a dog, ___ ?
5 They aren't coming, ___ ?
6 He does the shopping, ___ ?

SOUNDS

1 🔊 Listen and underline the stressed syllable.

1 England English   4 Japan Japanese
2 Germany German   5 Italy Italian
3 Canada Canadian   6 America American

Now say the words aloud. In which words does the stressed syllable change?

2 🔊 Listen and say these words aloud. Is the underlined sound /ɪ/ or /iː/? Put the words in two columns.

h<u>i</u>t h<u>ea</u>t b<u>i</u>t b<u>ea</u>t s<u>i</u>t s<u>ea</u>t b<u>i</u>n b<u>ea</u>n s<u>i</u>n s<u>ee</u>n g<u>i</u>n J<u>ea</u>n b<u>i</u>d b<u>ea</u>d h<u>i</u>d h<u>ee</u>d

3 🔊 Listen to three conversations. Which adjectives can you use to describe how each speaker sounds?

angry bored excited friendly frightened impatient

SPEAKING AND WRITING

1 Work in pairs.

Student A: Turn to page 98.
Student B: Student A is looking at a picture. Ask questions to find out what it shows.

2 Here are some sentences describing an English New Year's Eve party. Read them and put them in the right order. The first sentence is in the right position.

a We arrive at the party at about half past nine.
b We stand in a circle and sing 'Auld Lang Syne'.
c Just before midnight someone turns on the radio or television for the exact time.
d We kiss and wish each other 'Happy New Year'.
e We usually stay for another half-hour.
f We have a drink and something to eat.
g We listen to music, dance or talk.
h We go home to bed.
i At midnight we hear Big Ben in London.

3 Join the sentences together to make a paragraph. Use *after* and *before* clauses.

After arriving at the party at about half past nine, we ...

4 Write a paragraph describing how you usually spend New Year's Eve.

6 How we met

Past simple and past continuous

Josephine Wilson, **71, trained to be an actress but worked as a teacher for many years. She has three grown-up children and ten grandchildren. She lives with her dog in Los Angeles.**

Nguyen Van Tuan, **19, came from Vietnam to high school in the USA when he was sixteen. He now works as a pizza delivery boy in Long Beach, California. He has no contact with his family.**

VOCABULARY AND SPEAKING

1 What makes a good friend? Tick the qualities you like your friends to have.

| intelligence   sincerity   reliability |
| good looks   kindness   honesty |
| patience   single-mindedness |
| seriousness   idealism   optimism |
| openness   talent   confidence |
| ambition   informality   faithfulness |

Can you think of other qualities you like your friends to have?

I like my friends to have a sense of humour.

2 What are the adjectives formed from the qualities in the box?

Work in pairs and describe your ideal friends.

3 Work in pairs. Talk about when and where you met each other for the first time.

READING

1 You're going to read an interview with Josephine Wilson and Nguyen Van Tuan about how they met. First, read the caption to the photo and find out who they are.

2 Here are some questions about the interview. Before you read it, try to imagine the answers to the questions. Write full answers, but leave a blank if there is anything you cannot guess.

1 What was Josephine doing when he first saw her?
2 Where was Nguyen living when they met?
3 How was he feeling when they met?
4 What were they doing when they met?
5 Why was it difficult to hold conversations with him?
6 What was one of the first things Josephine did?
7 What did Josephine train to be?
8 Where did Josephine take Nguyen?
9 How was he coping with life in Los Angeles?
10 How often were they seeing each other at one point?

1 She was doing her shopping when he first saw her.

3 Work in pairs.

Student A: Turn to Communication activity 3 on page 98.
Student B: Turn to Communication activity 11 on page 101.

4 Work in pairs. Exchange information about the interviews you read, and compare and complete your answers to the questions in activity 2. Give each other as much information as possible about the interview you read.

GRAMMAR

Past simple
You use the past simple to talk about a past action or event that is finished.
Josephine Wilson trained as an actress.

Past continuous
You use the past continuous to talk about:
– something that was in progress at a specific time in the past.
He was feeling particularly lonely and was suffering from culture shock.
– something that was in progress at the time something else happened.
He was staying with Cathy Kelly when he first met Josephine Wilson.

1 Read the interview you read in *Reading* activity 3 again and underline all the verbs in the past simple and past continuous. Which past simple verbs are irregular?

2 Choose the correct verb form.

1 He *was working/worked* in a store, when she *was coming/came* in.
2 She *was doing/did* the shopping when he first *was seeing/saw* her.
3 While he *was living/lived* with Cathy Kelly, he *was meeting/met* Josephine.
4 Cathy Kelly *was having/had* a party when she *was introducing/introduced* Nguyen to her.

LISTENING AND WRITING

1 You're going to hear Sara, who is English and Joe, who is American, talking about their friendship. Here are some phrases they use.

just before last Christmas
similar interests and taste in music
at an end-of-term party
standing around with no one to talk to
talking to some friends
same sense of humour

What do you think the answers to these questions might be?

1 Where did they meet?
2 When did they meet?
3 What were they doing?
4 Why did they become friends?

2 Listen and check your answers to 1.

3 Work in pairs and think of any other details you heard.
Listen again and check.

4 Think of a special friend. Why did you become friends? Write notes in answer to these questions.

Where did you first meet?
When did you meet?
What were you doing?
How did you become friends?
What adjectives can you use to describe your friend?

5 Go round the class, asking and finding out about each other's special friends and make notes.

6 Write a paragraph about your special friend. Use your notes and the questions in 4 to help you.

The way things used to be

Used to and *would* + infinitive

HUNSTANTON

IT'S QUICKER BY RAIL

ILLUSTRATED GUIDE FROM SECRETARY ADVANCEMENT
ASSOCIATION OR ANY L·N·E·R OFFICE OR AGENCY

Investing in memories

My uncle had a moustache, a good job in the Civil Service and used to smoke forty cigarettes a day. But when he organised day trips for our family he used to behave like a boy. Today, seventy years later, we still talk about the wonderful trips to the sea our uncle used to organise when we were children.

He organised the trips very carefully. He used to buy the railway tickets and write special programmes long before the day arrived so we began to look forward to the trip. On the cover of the programme was the name of the place we were visiting and a humorous drawing of everyone in the family. There was even a lucky number on each programme and the winner didn't have to carry the bags on the way home.

All through the day he organised games and competitions. In his view, all the games had to be slightly anti-social. So if the programme said '4pm: Annual Ladies v Gentlemen football match', the match would always take place, even if the beach was very crowded. He organised treasure hunts, modelling competitions with seaweed, shells and bits of wood from the beach, and other events. There would be a special prize for the winner of every competition, usually an old sporting cup from a local junk shop.

He made sure that there were as many people on the trip as possible, and invited neighbours and their children as well to join the family for the day. The fun started as soon as we left home. Even the walk down to the station in the morning used to involve a game ('the first person to see a policeman gets a point').

One game we used to play in the car was called 'I know that lady'. One of us would choose someone walking along the street, and as we approached, the driver sounded the car horn, and everybody waved. The woman wouldn't understand why we were waving at her and would look puzzled.

He never thought money spent on a well-organised outing was wasted. When his wife complained about the cost of a family day out, he said, 'Look, it's not wasting money, it's investing in memories'.

Adapted from *Memories are made of this*
by David Randall, *The Observer*

VOCABULARY AND READING

1 You are going to read about someone's childhood memories of day trips to the seaside. Look at the words from the passage in the box and underline the verbs. Which nouns go with them?

> bags buy carry complain about cost family game invest in invite memories money neighbours play programmes spend tickets waste write

2 Read *Investing in memories* and tick (✓) the things that made the seaside trips so special for the author.

his uncle's job ☐
careful preparation ☐
carrying the bags ☐
the crowded beaches ☐
organised games and competitions ☐
prizes for the winners ☐
lots of people ☐
driving along in the car ☐
his uncle's wife ☐

3 Answer the questions and try to guess the meanings of the words or phrases.

behave – Does this mean something like *act* or *become?*
look forward to – Do they want the trip to the sea to happen or not?
seaweed – What kind of plants do you find on the beach?
junk shop – Is this a place where you can buy something new or old and second-hand?
puzzled – How would you feel if you didn't recognise people waving at you?

4 Work in pairs. Would you enjoy a trip to the sea like this?

GRAMMAR

> *Used to* and *would*
> You use **used to** + infinitive to talk about past habits, routines and states which are now finished. You often use it in narratives.
> *My uncle **used to smoke** 40 cigarettes a day.*
> *He **didn't use to smoke** cigars.*
> You often use **used to** to contrast past routine with present state.
> *I used to live in France. I don't live there now. I live in Britain.*
> You can also use **would** + infinitive to talk about past habits and routines, but not past states.
> *There **would be** a special prize for the winner.*

1 The following statements about *Investing in memories* are false. Read it again and correct them.

1 The writer's uncle used to have a beard.
2 He used to smoke cigars.
3 They used to have boring trips
4 They used to go to the mountains.
5 He used to invite only the family.
6 The fun used to start when they arrived.

2 Put a tick (✓) by the sentences where you can use *used to* or *would* and a cross (✗) where you can only use *used to.*

As a child, I *used to/would* spend my holidays in the mountains. We *used to/would* have a chalet near Geneva. We *used to/would* go there for a month in the summer, when it *used to/would* be sunny and for Christmas, when there *used to/would* be lots of snow.

SOUNDS

Look at your answers to *Grammar* activity 1 and listen to the stress and intonation when you correct someone.

Now say the sentences aloud. Try to use the same stress and intonation.

SPEAKING AND WRITING

1 Work in groups of two or three. Talk about how you used to spend your holidays in the past. Think about:

– where you used to go
– where you used to stay
– how you used to get there
– when you used to go
– who you used to go with
– what the weather used to be like

2 Work on your own. Write a description of how things used to be and how they are now. Use the expressions of time in **bold**.

Say how things used to be.
When I was a child, we used to go to our holiday home in Wales.

Say how things changed.
After a while, my brothers and sisters left home.

Say what happened in the end.
Eventually, my parents sold the house.

Say how things are today.
Now we go abroad for our holidays.

8 | Cold, lost, hungry and alone

Describing a sequence of events (2): *when, as soon as, as, while, just as, until*

VOCABULARY AND LISTENING

1 Work in pairs. Have you ever been cold, lost, hungry or alone? Say what happened and how you felt.

The last time I got lost was in London.   What happened?
Well, I arrived late one night and went to the centre, but...

2 You're going to hear Claudia, a journalist, talking about a visit to Russia on her own. Look at the words in the box and decide if she uses them to talk about when she was *cold, lost, hungry* or *alone*.

> afford banana beetroot bitter bread cabbage chilly crowds confused
> food foolish foreigner frightening freezing friend fur hat hospitable
> landmark lonely map pasty playground relieved sausage smile snow
> soldier stale stranger wind

3 🔲 Listen to part 1 of Claudia's story and tick (✓) the statement which describes her state.

| She was cold. ☐ | She was hungry. ☐ |
| She was lost. ☐ | She was alone. ☐ |

4 Number the events in the order they happened.

She forgot to ring her friend. ☐
A crowd of people surrounded her. ☐
Her German friend disappeared. ☐
They let her through. ☐
She was starting to panic. ☐
Someone tapped her on the shoulder. ☐
She was walking out into the hall. ☐
She felt lonely and rather foolish. ☐
She called Sergei. ☐
The theatre in St Petersburg sent a fax. ☐
She was getting out of the car. ☐
She was leaving London. ☐
They heard her speaking English. ☐
Three soldiers stopped her. ☐
She had to wait at reception. ☐
He was out. ☐

🔲 Now listen again and check your answers.

5 🔲 Listen to the part 2 of Claudia's story and decide which statement in 3 describes her state.

6 Decide if these statements about part 2 of the story are true or false.

1 As soon as she stayed with Sergei's family, she learnt some Russian.
2 While she was staying there, it was freezing cold.
3 She was enjoying the walk while she got lost.
4 She was standing by the side of the road while a car stopped.
5 The driver was going to work when he saw her.
6 Until the driver said 'taxi', she was very relieved.

🔲 Listen again and check your answers.

ЦПКиО

GRAMMAR

> *When, as soon as, as, while, just as, until*
>
> You use *when* and *as soon as* + past simple for actions which happen one after the other. The second verb is often in the past simple and is used for the action which interrupts the longer action.
>
> ***When*** *I saw a group of new arrivals, I wanted to join them.*
>
> ***As soon as*** *I arrived in Moscow, I intended to call my friend Sergei.*
>
> You can use *when, as* and *while* + past continuous for longer actions. The second verb is often in the past simple.
>
> ***As*** *we were queuing at passport control, a group of soldiers surrounded us.*
>
> You can use *just as* + past simple for shorter actions that happen at the same time. The second verb is often in the past simple.
>
> ***Just as*** *the plane was landing, the German businessman gave me a warning.*
>
> You can often use *when* instead of *while/as*.
>
> ***When*** *we got off the plane, I was feeling a bit nervous.*
>
> You use *until* to mean *up to the time when*.
>
> *Everything was fine* ***until*** *I came out of customs.*

1 Join pairs of sentences in *Vocabulary and listening* activity 4 with *when, as, while, as soon as* or *until*. There may be more than one possibility.

She forgot to ring her friend Sergei until she was leaving London.

2 Correct the false statements in *Vocabulary and listening* activity 6.

While she was staying with Sergei's family she learnt some Russian.

READING AND WRITING

1 Here are some questions about the last part of Claudia's story. Try to imagine the answers and write full answers to them. There will be some questions you can't answer completely, but just leave a blank.

1 Where did she go next?
2 Who was waiting for her at the station?
3 Why didn't the director recognise her?
4 How did she get to the theatre?
5 Who invited her to stay?
6 Where did Natasha live?
7 Who was sharing her flat?
8 Why couldn't Natasha afford to feed Claudia?
9 What did Claudia buy on the streets?

2 Turn to Communication activity 6 on page 99. Read the last part of Claudia's story and complete or correct the sentences you wrote in activity 1.

3 Write three paragraphs about Claudia's experiences in Russia. Use the linking words *when, as soon as, as, while, just as, until* if possible.

Paragraph 1: Write about when she was alone and cold. Use the sentences you wrote in *Grammar* activity 1 to help you.

Paragraph 2: Write about when she was lost. Use the true statements from *Vocabulary and listening* activity 6 and *Grammar* activity 2 to help you.

Paragraph 3: Write about when she was hungry. Use the sentences you completed or corrected in *Reading and Writing* activity 2 to help you.

9 Chocolate – like falling in love

Non-defining relative clauses: *who, which, where*

Chocolate first came from Central America.

The word *chocolate* comes from the Aztec language and is the only Aztec word we use regularly in English. The Aztecs made a greasy, bitter drink called *Xocoatl*, from cocoa beans mixed with cold water, spices and cornmeal.

The Aztecs used the cocoa bean as a form of money. According to H H Bancroft, who was a historian, 'four beans bought some vegetables, ten beans bought a woman and a slave cost 100'.

The explorer Cortes was the first person to bring chocolate to Europe. He presented it to the Spanish Royal Court in Madrid and served it with herbs and pepper. Soon it became very fashionable to drink it mixed with sugar and vanilla and drunk warm.

Coenrad van Houten, who was Dutch, was the first person to extract the cocoa butter from the cocoa bean in 1827.

In 1847, Joseph Fry, who lived in England, mixed the cocoa butter with other ingredients to make a solid chocolate bar.

Daniel Peter, who was a confectioner in Switzerland, invented milk chocolate in the 1870s. Henri Nestlé developed the process.

The cocoa tree originally comes from the Amazon rainforests. Brazil, West Africa and Ecuador now produce most of the 1.5 million-tonne world cocoa crop.

The Mexicans put chocolate in savoury dishes. They serve *mole*, which is a kind of chocolate sauce, with roast chicken.

It takes all the beans from one cocoa tree to make 500g of chocolate.

In Britain, people spend an average of 98p a week on chocolate. Women, who buy more than two-thirds of the chocolate, eat less than 40 per cent.

An average British person eats between 8.5 – 9.5 kg a year, except the Scots, who eat 30 per cent more.

Chocolate contains small amounts of the chemical phenylethylamine, which is also naturally present in the brain, and which gives us the same feeling as when we fall in love.

The world's largest chocolate model was a 10 m by 5 m representation of the Olympic Centre in Barcelona.

In 1980, the Swiss police arrested a young couple because they were trying to sell chocolate secrets to foreign powers. They offered the recipes for 40 different chocolates to the Soviet and Chinese embassies.

READING AND VOCABULARY

1 Do you like chocolate? Find out how many people in your class like chocolate. What kind of chocolate do they eat and drink most often?

2 Work in pairs. Which of these words do you associate with chocolate?

cocoa	sweet	savoury	bitter	bean
spices	herbs	pepper	sugar	milk
sauce	chemical	plain	healthy	
butter	bar	drink	solid	liquid
crop	greasy	grow	produce	
warm	cold			

Can you think of any other words you associate with chocolate? Does your partner agree with you?

3 Read the passage and find out why it's called *Chocolate – like falling in love*.

4 Match the following headings with each piece of information: *history, facts, interesting incidents.*

5 Work in pairs and decide where these non-defining relative clauses go in the passage.

1 ..., which people enjoy all over the world,
2 ..., which means 'bitter water',
3 ..., who first tried chocolate in 1519,
4 ..., who was Peter's colleague,
5 ..., which weighed nearly 200 kg,
6 ..., where they held the Olympic Games in 1992

GRAMMAR

> **Non-defining relative clauses: *who, which, where***
>
> You use a non-defining relative clause with *who*, *which* or *where* when you add extra information to a sentence. You use *who* for people.
> *Van Houten, **who** was Dutch, was the first person to extract chocolate from cocoa.*
> You use *which* for things.
> *Chocolate, **which** people enjoy all over the world, first came from Central America.*
> You use *where* for places. Remember that *where* is an object pronoun and must be followed by a subject.
> *Chocolate first came from Central America, **where** the Aztecs lived.*
> You put the non-defining relative clause immediately after the person or thing it refers to. The relative pronoun replaces the second noun or pronoun.
> *Cortes was the first person to bring chocolate to Europe. **He** was an explorer.*
> *= Cortes, who was an explorer, was the first person to bring chocolate to Europe.*
> You separate the relative clause from the main clause with commas, or a comma and a full stop if it is at the end of the sentence.

1 Look at *Chocolate – like falling in love* and underline the non-defining relative clauses.

2 Complete these sentences with *who, which* or *where*.

1 The word *chocolate*, ___ comes from the Aztec language, is the only Aztec word in English.
2 The first person to bring chocolate to Europe was Cortes, ___ was an explorer.
3 The Aztecs made a drink from cocoa beans ___ was called *Xocoatl*.
4 Henri Nestlé, ___ was Swiss, developed the process of making milk chocolate.
5 Chocolate, ___ contains a special chemical, makes us feel as if we are falling in love.
6 In Mexico they serve chicken with *mole*, ___ is a kind of chocolate sauce.
7 People in Britain spend 98p a week on chocolate except in Scotland, ___ they spend more.
8 The police arrested a couple in Switzerland, ___ they were trying to sell chocolate secrets.

3 Rewrite these sentences with *who* or *which*. There are two possible answers for each sentence, depending on which sentence you think contains the extra information.

1 Champagne is one of the most expensive drinks in the world. It comes from France.
2 The Brazilians export the most coffee in the world. They produce a million tonnes a year.
3 Dr John Pemberton invented Coca Cola. He lived in Atlanta, USA.
4 The avocado pear contains the most calories of any fruit. It has more protein than milk.
5 The Incas discovered popcorn. They lived in South America in the fifteenth century.
6 The durian fruit has a disgusting taste and smell. It is considered by some people to be a delicacy.

SPEAKING AND WRITING

1 Work in pairs. Make a list of three well-known things, people, or places.

Tea,...

2 Ask other students for two pieces of information about the things, people and places in your list. Think about history, present day facts and interesting incidents.

Tea – grows in India and China, national drink of Britain,...

3 Write a description of things, people or places, linking the extra information with *who, which* or *where*. Try to add as much extra information as possible.

Tea, which grows in India and China, is the national drink of Britain.

10 | *What did I do?*

Verbs with two objects; complaining and apologising; making requests

READING AND VOCABULARY

1 When we're with people from different countries, we sometimes have misunderstandings, or make mistakes. Read *What did I do?* and decide what each misunderstanding or mistake is.

2 Work in pairs and check your answers to 1. Would the situations cause a misunderstanding or mistake in your country?

3 Here are some of the words from the passage. First, decide what part of speech they are – noun, verb or adjective. Then match them with their meanings in the context of the passage.

> wallet hurt customer colleague
> waiter mean bunch
> meal crowded

1 an occasion when people eat together
2 someone who buys something
3 full of people
4 a group of picked flowers
5 someone you work with
6 something men keep banknotes and credit cards in
7 a man who serves you in a restaurant
8 unkind, too careful about money
9 emotionally upset

When I was at university in England, my English tutor invited a group of us to her home. I didn't want to make any mistakes, such as staying too late. So when she brought us a drink before we began the meal, I said, 'Thank you for inviting us to your home and for inviting us to dinner. Could you tell me when we can leave?' She laughed and said, 'So, you can't wait to leave?'. **Lu, China**

I was visiting Germany for the first time and I received an invitation to visit my most important customer in her house. I decided to take her a beautiful bunch of twelve red roses and her husband a bottle of wine. I gave her the flowers, but she just looked embarrassed. **Douglas, Scotland**

A British colleague invited me to join his friends after work. We went to a pub where he bought me a drink and he suggested a meal in a restaurant. At the end of the meal, I was very surprised to see everyone take out their wallets to pay the waiter. My friend expected me to pay as well, but I feel it was very mean of him not to pay for me as he invited me. **Kenji, Japan**

I was sitting in a bus in Bristol when an elderly lady got on the bus. It was crowded and there weren't any seats. A middle-aged man said very loudly, 'Would you offer the lady your seat, please?' Why didn't he give her his seat? **Carlos, Spain**

I've only recently arrived in the USA and don't have many friends so I was pleased to meet a really nice American in the college cafeteria the other week. We had a long conversation, she told me the story of her life, she showed me photos of her family, and she left me her address. The following week I saw her, but although she smiled and said 'Hi!' in a friendly way, she went and sat with other friends. I feel very hurt. Does she expect me to call on her? I feel I need an invitation. **Hana, Lebanon**

LISTENING

1 🔲 Listen to someone explaining what the misunderstandings or mistakes are. Did you guess correctly in *Reading and vocabulary* activity 1?

2 Work in pairs and try to remember in detail the explanation of each misunderstanding. Is there anything that surprises you?

🔲 Listen again and check.

GRAMMAR AND FUNCTIONS

> **Verbs with two objects**
> **Many verbs can have two objects, a direct object and an indirect object. You usually put the indirect object, which often refers to a person, before the direct object.**
> *He bought **me** a drink.*
> **You can put the indirect object after the direct object with a preposition, usually *to* or *for*.**
> *He bought a drink **for me**.*
> **Here are some other common verbs which can have two objects: *cost, give, lend, make, order, pass, pay, promise, read, sell, send, teach.***
>
> | Complaining | Apologising |
> |---|---|
> | *I'm sorry but it's too noisy here.* | *I'm (very) sorry. I really am sorry.* |
> | *I'm afraid it's too noisy here.* | *Excuse me. I'm awfully sorry.* |
>
> | Responding to apologies | Making requests |
> |---|---|
> | *That's all right/OK.* | *Could you offer the lady your seat?* |
> | *Don't worry about it.* | *Would you mind giving the lady your seat?* |
> | *Never mind. It doesn't matter.* | |

1 Read *What did I do?* again and find at least six verbs which can take two objects. Rewrite the sentences using *to* or *for*.

2 Rewrite the sentences without using *to* or *for*.

1 She took some flowers for her friends.
2 Can you give the keys to her?
3 I'm bringing a nice cake for you.
4 Could you send a cheque to them?
5 They'll order a taxi for you at reception.
6 Write a postcard to them when you get there.

1 She took her friends some flowers.

3 Write sentences describing what you say in these situations.

1 Someone is smoking too much.
2 You've forgotten your wallet.
3 The television next door is too loud.
4 Your friend is driving too fast.
5 Your new watch doesn't work.
6 You're late for a dinner party.

SOUNDS

🔲 You are going to hear each sentence below twice, first as a complaint and then as an apology. Listen to the way the speaker complains or apologises.

1 I'm sorry I didn't hear you.
2 I'm afraid I'm going to be late.
3 I'm sorry but it's too late.
4 I'm afraid I can't talk now.
5 I'm very sorry but I can't do that.
6 I'm afraid I haven't got time now.

Now say the sentences aloud. Try to make a difference between complaining and apologising.

SPEAKING

1 Work in pairs. What do you do if there is a misunderstanding or if you realise you've made a mistake?

– apologise immediately
– make an excuse
– feel embarrassed
– make sure you don't do it again
– hope that people will forgive you
– blame someone else
– change the subject
– make a joke
– say nothing
– explain that you do things differently in your country

2 Talk about situations when there was a misunderstanding or you made a mistake. Do you think it mattered very much?

Progress check **6–10**

VOCABULARY

1 Some adjectives have 'strong' equivalents, adjectives which have a similar but stronger meaning. For example:

good – excellent, hot – boiling, big – huge

Here are some of the adjectives which have already appeared in *Reward* Intermediate.

annoying angry cold frightening funny interesting pleased silly surprised

Match them with their 'strong' equivalents in the box below.

> astonished delighted fascinating freezing furious
> hilarious infuriating ridiculous terrifying

2 You can add the prefixes *un-, im-, in-, dis-, ir-*, and *il-* to many adjectives to give the opposite meaning.

uninteresting impossible inexpensive dishonest irregular illogical

The most common of these prefixes are *un-, im-* and *in-*. Write the opposite of these adjectives by adding *un-, im-* or *in-*.

certain clear patient correct friendly kind complete happy important modest perfect

3 Some English words sound the same but have different spelling and meaning. These are called homophones, for example:

meat – meet, weak – week, new – knew, weight – wait, fare – fair

Write down homophones for these words. You can find them in lessons 6–10.

been beech by bred plane waist right

4 Look at the vocabulary boxes for lessons 6–10 again. Choose words which are useful to you and write them in your *Wordbank*.

GRAMMAR

1 Choose the correct verb form.

An American who *visited/was visiting* Russia, *wanted/was wanting* to go on a wild bear hunt and *paid/was paying* a lot of money for the sport. The travel agent *took/was taking* the American to Moscow's Perdelino Forest. Suddenly, he *saw/was seeing* a bear, and *decided/was deciding* to get closer because he *hoped/was hoping* to shoot it. A postman, who *rode/was riding* past on his bicycle, *fell/was falling* off in surprise when he *spotted/was spotting* the bear. The bear *came/was coming* over to the bicycle, *picked/was picking* it up and *rode/was riding* off. In fact, it wasn't a wild bear at all. It *performed/was performing* at the local circus. The bear *escaped/was escaping*, the postman *lost/was losing* his bicycle and the American *asked/was asking* for his money back.

2 Choose the correct verb form.

Howard Hughes, the American film producer, *lived/used to live* for fifteen years completely on his own. He *was spending/used to spend* all day lying on his bed watching films. He *would/used to* hate touching anything that *wasn't/didn't use to be* sterile. He *was living/used to live* on tinned chicken soup for weeks, and then *was changing/used to change* to a diet of ice cream. When he *died/used to die*, he *was/used to be* a billionaire.

3 Complete the sentences with *as, just as, when, while, as soon as, until*. There may be more than one possible answer.

1 ___ the traffic stopped, we ran across the road.
2 He stayed in bed ___ he got better.
3 ___ I was living in London, I spent a lot of time in the museums.
4 ___ I saw her, I waved to her.
5 ___ I was leaving, the phone rang.
6 She waited by the door ___ someone opened it.

4 Join the two sentences together with a non-defining relative clause.

1 Venice is in the north of Italy. It stands on 118 islands.
2 The London Underground has 400 km of tunnels. It is the longest in the world.
3 A Dutchman bought Manhattan island in 1626. He only paid about $24.
4 Barbara Cartland is the world's most popular writer. She has sold about 400 million copies of her romantic novels.
5 The oldest city in the world is Jericho. It had a population of 3000 people in 7,800 BC.
6 Women live longer than men. They have an average life of 77 years.

5 Rewrite these sentences without using *to* or *for*.

1 I gave a present to her.
2 She brought a bottle of wine for me.
3 He passed the dish to her.
4 We sent a postcard to her.
5 My father taught French to children.
6 I showed my passport to the police.

SOUNDS

1 In connected speech, it's usual to link words. If you don't, it makes you sound very formal. Listen to some of the types of sounds which you link. As you listen, say the sentences aloud.

– consonant + vowel
She had a German accent.

– vowel + vowel using /j/, /w/ or /r/
It'll be interesting. Please go in. It flies via Iceland.

– same or similar consonants
We used to eat toast at breakfast.

2 Listen and mark the word links.

My uncle had a moustache, a good job in the Civil Service and used to smoke forty cigarettes a day. But when he organised day trips for our family he used to behave like a boy. Today, seventy years later, we still talk about the wonderful trips to the sea we used to have as children, and which our uncle used to organise.

WRITING AND SPEAKING

1 In the story below there are some words which are not necessary for the general sense of the passage. For example, in the first sentence, *close, often, beautiful* and *weekly* are not necessary. Read the story and cross out any other 'unnecessary' words. You cannot cross out two or more words together.

A close friend of my mother lives in the country in Yorkshire, and she often goes to the beautiful town of Harrogate to do her weekly shopping. After doing the shopping she usually has tea in a small tea shop. One afternoon she was looking forward to having tea and she went to her usual tea shop, but it was crowded with people from the antiques fair. The waitress was rather embarrassed that there was no room for a regular customer, but she offered her a place at a small table, sharing with a middle-aged man. The lady was disappointed but wanted a cup of tea very much, so she agreed. The waitress showed her a table by the front window, where the man was sitting. The man smiled politely then returned to his paperback book. After a few minutes he got up and left.

The lady was drinking her tea when she noticed that there was a slice of fruit cake on the man's plate. She looked around carefully but there was no sign of him. She was feeling extremely hungry and it seemed a dreadful shame to waste it, so she picked it up and ate it. Just as she was finishing the cake, the man reappeared and returned to the table.

2 The following words can go in the passage: *cold*, *extremely*, *young*. They can go in the following positions: *One **cold** afternoon, it was **extremely** crowded, The **young** waitress.*

Write six more words which can go in the passage.

3 Work in pairs. Show your words to your partner and decide where his/her words can go.

4 Work in groups of three or four. What's the most embarrassing incident you can remember? Tell each other about it... if you can!

Present perfect simple (1): *already, yet, still*

READING

Read and answer the questionnaire.

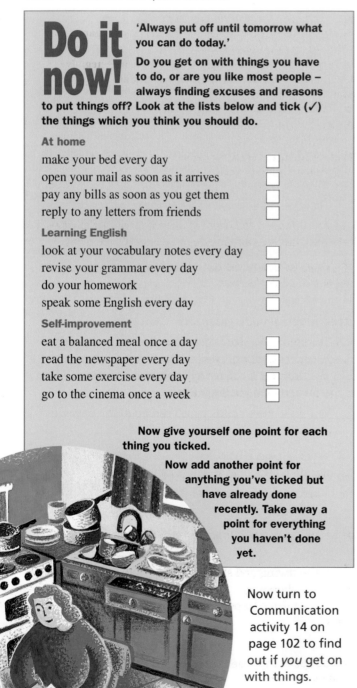

Do it now!

'Always put off until tomorrow what you can do today.'

Do you get on with things you have to do, or are you like most people – always finding excuses and reasons to put things off? Look at the lists below and tick (✓) the things which you think you should do.

At home

make your bed every day ☐
open your mail as soon as it arrives ☐
pay any bills as soon as you get them ☐
reply to any letters from friends ☐

Learning English

look at your vocabulary notes every day ☐
revise your grammar every day ☐
do your homework ☐
speak some English every day ☐

Self-improvement

eat a balanced meal once a day ☐
read the newspaper every day ☐
take some exercise every day ☐
go to the cinema once a week ☐

Now give yourself one point for each thing you ticked.

Now add another point for anything you've ticked but have already done recently. Take away a point for everything you haven't done yet.

Now turn to Communication activity 14 on page 102 to find out if *you* get on with things.

LISTENING AND VOCABULARY

1 You're going to hear two people talking about what they've got to do. Read their conversation. Do you notice anything strange?

IAN   Hi, Kate. How are you getting on?

KATE   Hey, what are you doing here? I didn't expect you until later.

IAN   Well, I've already finished everything I had to do at work, so I thought I'd come back and get in the way. I know you've had a relaxing day.

KATE   I've had an extremely busy day, and it hasn't finished yet. Did you remember we've got Paul and Hannah for dinner tonight?

IAN   Yes, I did. Have you got everything you need?

KATE   Well, I've already done the shopping, but I haven't sold the wine yet.

IAN   I'll do that. Have you dropped off the children yet?

KATE   No, I haven't done that yet. They're still at school.

IAN   OK. Well, I'll get them when I buy the wine.

KATE   No, they've got their sports club, so they'll be there until about six. You're more important here. Have you taken the car to the garage yet? It should be ready by now.

IAN   No, I haven't. I'll go to the garage later. Now, what's next?

KATE   Well, the kitchen is dirty, but you could check the bathroom for me.

IAN   OK. Have you tidied up the pile of newspapers in the front room?

KATE   No, they're still standing on the table.

IAN   I'll bring in the rubbish as well.

KATE   I've done that already. Now, I haven't started cooking yet, and it's getting late. When you've got the wine, can you clear the table? Here's the table cloth and knives and forks. And damage the television. It still doesn't work. And unwrap your mother's birthday present. It still needs wrapping paper. And turn off the heating. It's getting quite cold in here.

IAN   Oh dear, it's one of those days.

2 🔊 Listen and underline anything in the conversation that is different from what you hear.

3 Replace the words you underlined with words from the box below. There is one extra word.

> busy buy clean collect
> help lay lie mend pick up
> take out throw away tidy
> turn on wrap up

Collect the car.
Buy the wine.
Check the bathroom
Take out the rubbish

4 Look at the list of things to do. Listen and put a tick (✓) by the things Kate has already done and a cross (✗) by the things she hasn't done yet

do the shopping ☐
take out the rubbish ☐
lay the table ☐
buy the wine ☐
collect the car ☐
start cooking ☐
clean the kitchen ☐
throw away the newspapers ☐
pick up the children ☐
check the bathroom ☐
wrap up the birthday present ☐
mend the television ☐
turn on the heating ☐

5 🔊 Listen again and check your answers to 3 and 4.

GRAMMAR

> Present perfect simple (1): *already, yet, still*
>
> **You can use the present perfect simple to talk about:**
> **– something which happened in an indefinite time in the past, such as an experience, with *ever* and *never*.**
> *Have you **ever** read an English newspaper?*
> **– a past action which has a result in the present, such as a change.**
> *She's **done** the cooking.*
> **You can use *already* with the present perfect simple to suggest *by now* or *sooner than expected*. It's often used for emphasis and goes at the end of the clause.**
> *I'll bring in the rubbish.   No, it's OK, I've done that **already**.*
> **You can also put *already* between the auxiliary verb and the past participle. You don't often use *already* in questions and negatives.**
> **You can use *yet* with the present perfect simple in questions and negatives. You use it to talk about an action which is expected. You usually put *yet* at the end of the sentence.**
> *Have you picked up the children **yet**? I haven't bought the wine **yet**.*
> **You use *still* to emphasise an action which is continuing.**
> *The children are **still** at school.*
> **You usually put *still* before the main verb, but after *be* or an auxiliary verb. In negatives it goes before the auxiliary verb.**
> *He **still** hasn't done it.*

1 Work in pairs. Look at your answers to *Listening and vocabulary* activity 4. Ask and say what Kate and Ian have already done or haven't done yet.

Has Kate done the shopping yet? - Yes, she has.
Has she bought the wine yet? - No, she hasn't.

2 Complete these sentences with *already, yet* or *still*.

It's half past eight in the morning, and Jack should be at school by nine but he's ___ in bed. He's ___ had breakfast because he brought it back to bed. He hasn't packed his bags ___ , and he is ___ trying to make up his mind what to wear. He's ___ been late for school three times this week.

SPEAKING

1 Work in pairs. Talk about your answers to the questionnaire. Do you think your score describes you accurately?

2 Look at the activities below. How much time do you usually spend doing them each week?

– earning money
– eating and drinking
– doing housework
– doing leisure activities
– resting and sleeping
– shopping
– talking to people
– washing and dressing
– travelling
– studying

3 How much time have you spent this week on the activities in 2? Is this week typical? Why/why not?

12 *London calling*

Present perfect continuous (1) for asking and saying how long

VOCABULARY AND LISTENING

1 Look at the words in the box. In which type of news story would you expect to hear them? Choose from the list of types of news stories below.

> air force army attack battle bomb border bullet club currency
> defeat demonstration election fire gale gun hijack hostage judge
> jury kidnap lawyer left wing lightning lose march match missile
> navy officer party player president prices prime minister rain
> right wing soldier stadium team terrorist thunderstorm troops
> victory war weapon win

	1	2	3	4	5	6
Arts and culture						
Business and economics						
Crime						
Disaster						
Famous people						
Foreign Affairs						
Government						
Law						
Leisure interests						
Military						
Politics						
Religion						
Social affairs						
Sport						
Traffic and transport						
Weather						

2 🔊 You're going to hear the six o'clock news, which has six main news stories. Listen and decide what type of news story each one is. Put a tick (✓) in the appropriate boxes in the chart. Some stories may be of more than one type.

3 Work in pairs. Try to remember each news story.

🔊 Now listen again and check.

GRAMMAR

> **Present perfect continuous (1)**
> **You use the present perfect continuous to talk about actions and events which began in the past, continue up to the present and may or may not continue into the future. You use** *for* **to talk about the length of time.**
> *I've been learning English for three years.*
> **(I started learning English three years ago, and I'm still learning.)**
> **You use** *since* **to say when the action or event began.**
> *He's been living here* **since** *1989.*
> **You can also use the present perfect continuous to talk about actions and events which have been in progress up to the recent past, especially when the results are present.**
> *It's been raining.* **(It's not raining now, but the ground is still wet.)**
> **You don't usually use certain verbs (***know, like, think, want* **etc) in the continuous form.**
> **You form the present perfect continuous with** *has/have been* **+ -ing. You usually use the contracted form** *'ve* **or** *'s.*
> *I've been working. She's been teaching. They've been dancing.*
>
> **Questions**
> *What have you been doing? Have you been working?*

1 Work in pairs. These sentences about the news stories you heard are wrong. Write them out again and correct the wrong information.

1 The hostages have been sitting in the plane since Wednesday.
2 The people in the Balkans have been waiting for food for eight months.
3 Traffic in west London has been growing since early yesterday evening.
4 The demonstrators have been marching through Manchester for several days.
5 The fire on the oil rig has been burning since Monday.
6 It has been raining in Scotland for seven hours.

2 Write sentences using the present perfect continuous.

1 The sun came out three hours ago. It's still shining.
2 He started work in 1956. He's still working for the same company.
3 She started to live with him in 1965. She still lives with him.
4 We had our first holiday in Majorca ten years ago. We still go there.
5 I started learning Italian in 1987. I'm still learning it.
6 Prices started rising three months ago. They are still rising.

1 The sun has been shining for three hours.

3 Write questions using *how long* and the present perfect continuous.

1 do this lesson? 2 use this book? 3 learn English? 4 live in your town?

4 Ask other students your questions and note down their answers. Answer their questions using *since*.

5 Write sentences about other students using *for*.

SOUNDS

1 🔲 Listen to the answers to *Grammar* activity 1. Notice how the speaker stresses words to correct the information.

Now say the sentences aloud.

2 Read this news story and underline the words you think the speaker will stress.

Hijackers are still holding twenty-three passengers in a plane at Manchester airport. They hijacked the flight from London to Glasgow last Thursday. The hostages have now been sitting in the plane without food or water for three days.

🔲 Listen and check. Read the news story aloud.

WRITING AND SPEAKING

1 🔲 Listen to one of the news stories again.

Now turn to Communication activity 7 on page 99.

2 Decide if you agree with these statements.

Most people are only interested in bad news.
There are certain circumstances, such as war, when the government should control the news.
There is too much news about people and personalities and not enough about politics and current affairs.
People are only interested in news which happens near them or which affects them economically.

3 Work in groups of three or four and discuss your answers to activity 2.

13 *Fictional heroes never die*

**Present perfect simple (2) and
present perfect continuous (2)**

READING

1 Look at the pictures on these
pages. Do you recognise these
characters? Which ones do you
know? Here are the names of the
characters in English. Match the
names with the characters.

Kermit (1957–)
Bugs Bunny (1937–)
James Bond (1953–)
Batman (1939–)
Charlie Brown (1950–)

What are their names in your
country? Can you think of other
fictional heroes like them?

Tintin, Asterix...

2 Read *Fictional heroes never die*
and match the names in 1 with
the paragraphs.

3 Answer the questions.

1 How long has Batman been
saving Gotham City from evil?
2 How long has Charlie Brown
been living in a small American
town?
3 How long has Bugs Bunny
been eating carrots in public?
4 How long has Kermit been
singing and dancing on TV?
5 How many Batman TV series
has the team made?
6 How many Charlie Brown
films have there been?
7 How many cartoon
films has Bugs Bunny
appeared in?
8 How many Muppet
shows have people
seen?

Fictional heroes never die

Real heroes are only human.
They live and die like the rest of us.
But fictional heroes are different.
They began life many years ago
and they will live on in fiction in the
future. Here are some favourite
twentieth-century fictional heroes.

1 The spy 007 has been working for the
British Secret Service since 1953 when
Ian Fleming first wrote about him in *Casino
Royale*. Fleming is now dead, but the spy still
lives on. The first film was *Dr No* in 1962 and
since then he's appeared in over twenty films.
The Cold War has finished and Bond's
enemies are not the Russians now, but there
are still plenty of bad people out there!
Columbia, the film producer, recently finished
filming his latest adventure.

2 He first saved Gotham City
from evil, with the help of
Robin, in a comic book in 1939 and
has been fighting crime for over
sixty years. The team had made
three TV series, as well as films and
novels. The most popular TV series
was in the 1960s, and our hero has
recently appeared in films. He will
live on as long as there are
criminals like the Penguin,
Catwoman and the Joker in Gotham
City.

3 This character has been living with his
pet dog Snoopy in a small American
town since 1950, when he first appeared in
the strip cartoon *Peanuts*. Over forty years
later, he still experiences all the happiness
and frustration of a typical boy, playing
baseball with his friends and going to school.
His first appearance in a TV film was in 1965
and so far he has made three cinema films.
His appeal is as strong as ever.

4 The talking rabbit first asked 'What's Up
Doc?' in 1937 and has been eating
carrots in public ever since. He has appeared
in comic books, newspaper strips and above
all, hundreds of cartoon films. He will be
popular for as long as people laugh at rabbits.

5 The green frog has been singing and dancing on TV since 1957, although it was
only in 1967 that he became famous with the Muppets. Since then, 235 million
people in 100 countries have seen hundreds of Muppet shows. There have also been
three films. Success has not changed him; he remains
exactly the same colour.

GRAMMAR

> **Present perfect simple and present perfect continuous**
>
> **It is sometimes difficult to know when to use the present perfect simple and the present perfect continuous.**
>
> **You use the present perfect simple:**
>
> **– when the action is finished.**
>
> **– to say what has been completed in a period of time.**
>
> *James Bond has appeared in over twenty films.*
>
> **(The appearances are now in the past; we're interested in how many films.)**
>
> **You use the present perfect continuous:**
>
> **– when the action may or may not be finished.**
>
> **– to say how long something has been in progress.**
>
> *He's been working for the British Secret Service since 1953.*
>
> **(He may continue to work for them in the future; we're interested in how long he's been a spy.)**

1 Look back at the passage and underline the verbs in the present perfect continuous.

2 Write full answers to questions 1–4 in *Reading* activity 3. Use the present perfect continuous, and *for* and *since* in turn.

3 Write full answers to questions 5–8 in *Reading* activity 3. Use the present perfect simple.

4 Choose the best tense.

1 I*'ve worked/been working* for several hours and I still *haven't finished/been finishing*.

2 She*'s waited/been waiting* for thirty minutes and the bus still *hasn't arrived/been arriving*.

3 I*'ve read/been reading* while she*'s seen/been seeing* the doctor.

4 He*'s talked/been talking* to his girlfriend for over an hour. That's the sixth phone call he*'s made/been making* this evening.

VOCABULARY

Here are some words from the passage. Check you know what they mean. Try to remember the sentence in the passage in which you saw them.

appear   baseball   cartoon   comic   criminal   enemy   fictional   frog   hero   laugh   pet   popular   rabbit   series   spy   success   team   typical   war

Look back at the passage and check.

SPEAKING AND WRITING

1 Think about some important events and dates in your past which relate to your life now.

1972 – born in Granada, 1989 – moved to London, 1993 – went to university, 1994 – started work, bought a car

Now think of two or three fictional events.

1995 – married Bugs Bunny,...

2 Work in pairs. Find out about important events and dates in your partner's life and make notes.

Mercedes – born 1972, has been living in London since 1989, has been working since 1994...

3 Use the biographies in *Fictional heroes never die* to write a biography of your partner. Say how long he or she has been doing things, how many things he or she has done, and from when to when he or she did things. Include the fictional information.

Mercedes Sanchez 1972–
Mercedes was born in 1972 in Granada. She has been living in London since 1989. She has been married to Bugs Bunny for a number of years.

4 Show your biography to another student. He or she must try and find the fictional information.

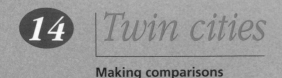 

14 | *Twin cities*

Making comparisons

VOCABULARY AND LISTENING

1 Underline the adjectives in the box below. Are there any which can also be nouns?

art gallery   bridge   busy   castle
cathedral   cemetery
coffee-house   concert hall
cosmopolitan   crowded
dangerous   district   dirty   hilly
industrial   lane   lively   market
museum   noisy   old-fashioned
palace   park   picturesque   quay
river   romantic   skyscraper
sleepy   smart   square   street
theatre   tram   unique   wealthy

2 Think of your favourite town or city. Which of these adjectives can you use to describe it?

3 What special features is your favourite city famous for? Use the nouns in the box to help you. Are there any adjectives which go with them?

Lyon is famous for its medieval district and its picturesque rivers.

4 🔲 You're going to hear Richard, who lives in Prague, comparing the city with London. Look at the photo of Prague and decide which adjectives in the box he might use to describe it.

Now listen to Richard.
Did he use the adjectives from the box that you chose?

5 Work in pairs. Which of the following aspects of Prague does Richard talk about? What does he say about them?

age   population   industry   climate   prices   architecture   transport
cleanliness   safety   atmosphere   culture   entertainment

🔲 Listen again and check.

GRAMMAR AND FUNCTIONS

Making comparisons
You can make comparisons in the following ways:

– **comparative adjective + *than***	*Prague is **cheaper than** London.*
– ***more/less* + comparative adjective + *than***	
	*Rome is **more beautiful than** London.*
	*London is **less dangerous than** New York.*
– ***more* + countable/uncountable noun + *than***	
	*Tokyo has **more inhabitants than** Madrid.*
	*Berlin has **more rain than** Rome.*
– ***fewer* + countable noun + *than***	*Madrid has **fewer tourists than** Paris.*
– ***less* + uncountable noun + *than***	*Paris has **less industry than** Milan.*
– ***as many* + countable noun + *as***	*Milan has **as many people as** Barcelona.*
– ***as much* + uncountable noun + *as***	*Barcelona has **as much rain as** Rome.*
– ***as* + adjective + *as***	*Rome is **as beautiful as** Prague.*

1 Look back at *Vocabulary and listening* activity 1 and check you can form the comparative and superlative forms of the adjectives.

2 Write sentences comparing the aspects of London and Prague in *Vocabulary and listening* activity 4.

Fewer people live in Prague than in London.

READING AND WRITING

1 Read the passage about Prague. Which of the aspects in *Vocabulary and listening* activity 5 does it mention?

2 There is a mixture of fact and opinion in the passage. For example, it is a fact that there are Gothic and baroque buildings, but an opinion that the clock in Old Town Square is *wonderful.*

Read the passage again and write down:

– five facts about Prague.

over a thousand years old,...

– five opinions about Prague.

one of the most memorable sights in Europe,...

3 Think of a town which you can 'twin' with Prague. Write notes about the features of the two towns. It should have as many similar aspects as possible.

London: similar age, culture

Are there any major differences?

London: no trams, different architecture

4 Write a description of the twin cities, giving your reasons why they are similar. Use the notes you made in 2 and 3 and the linking words in **bold**.

Give the main reason.	***The main reason why*** *Prague and London are similar is their age.*
Give other reasons.	***Another reason is that***...
Describe similarities.	***Both*** *Prague* ***and*** *London have a lot of old buildings.* ***Neither*** *Prague* ***nor*** *London has much heavy industry.*
Describe differences.	*Prague has hot summers* ***while*** *London can be colder. Prague is very romantic* ***whereas*** *London is more reserved.*
Stress similarities.	*But actually, they are* ***both*** *very cosmopolitan cities.*

Nowhere quite like Prague

Is there any city in Europe, or elsewhere, like Prague? There has been a city here for over a thousand years, and now 1,250,000 people live here. It is most famous for its Gothic and baroque buildings. Old Town Square, with its wonderful clock, the Charles Bridge, and Prague Castle on the hill above the river are just a few of Prague's famous attractions.

Getting around Prague is easy by tram or underground but it's also a pedestrian's dream because much of the old quarter and many of the streets and lanes have little or no traffic. The medieval centre is Prague Castle and St Vitus Cathedral. An evening view of these illuminated landmarks is one of the most memorable sights in Europe. Wenceslas Square is in the heart of modern Prague.

Visiting Prague today, you immediately notice the lively atmosphere. The city can be crowded during the hot summer months, but it is a delight to visit at any time of the year, even in the snowy cold of winter. In fact, tourism makes the largest contribution to Prague's economy. Classical concerts take place all through the year, though the biggest event is the Prague spring International Music Festival in May and early June. Theatre also has a special place in the life of the city.

The suburbs are like many in Eastern Europe with tall skyscrapers and some light industry, but you're very quickly in the sleepy villages and gentle hills of Bohemia.

Many people say Prague reminds them of Vienna or Budapest. But in fact, Prague is unique. There's nowhere quite like it.

15 | *I couldn't live without it*

Adjectives (2): order of adjectives

How many possessions do you have? Are your possessions very important to you or can you live without them? We asked a few people about their favourite things.

Is there anything you couldn't live without?
'My wonderful, brand-new mountain bike. I got it for Christmas and I keep it in my bedroom, although it's difficult to get upstairs every night.'
Nick, Abingdon

What's the first object you like to see when you get up in the morning? 'The thing I like best is one of our large, white, French coffee cups full of steaming coffee and hot milk.'
Josephine, Chichester

What's the most useful thing you have at home? 'I'm incredibly lazy, so the most useful thing is the remote control which I can use to change channels on the TV without having to get up out of my armchair.' *Lesley, Matlock*

What non-essential object do you usually take with you when you go out? 'I have a lovely, old, silver fountain pen with my initials on it, which I take everywhere. I love it because it feels very solid and reassuring.' *Barry, St David's*

What's the most valuable thing you own? 'A beautiful, antique, gold engagement ring with rubies. It was my grandmother's ring but it's mine now. It must be at least a hundred years old because I think it was her mother's before she had it.' *Laura, Newcastle*

What's the oldest thing you own? 'We've got two charming, eighteenth-century prints of Cambridge, which we're very fond of, which hang side-by-side on the living room wall in a couple of rather nice rectangular frames. We were both at the university and we bought them before we met. In fact, mine is older than my husband's, but his is probably worth more.' *Helen, Edinburgh*

What was the last present you bought for yourself? 'I don't really buy myself presents very often, only essentials, really. But I did buy a heavy, glass ashtray when I was on holiday in Spain last year. We don't smoke, but we have several friends who do and we always have to hunt for an ashtray when they come to visit.' *Diana, Macclesfield*

What would you save if there was a fire in your home? 'The most valuable toys I have are my collection of plastic model airplanes, and I'd try to grab them all if there was a fire. I've also got a small, black book which is full of information about nature and birds to watch out for, and things like that, and a whole series of really funny jokes.' *Simon, Uffington*

VOCABULARY AND READING

1 Look at the words in the box. How many compound nouns are there?

> alarm clock antique armchair
> ashtray bicycle china coffee cup
> cotton diamond engagement ring
> fountain pen glass gold
> heavy lamp metal necklace
> painting paper photo frame
> picnic basket plastic
> porcelain rectangular round
> rubber sofa square torch vase
> violin wool wood wristwatch

Which compound nouns are:

– one word formed from two nouns? *ashtray*
– two words formed from two nouns? *alarm clock*

2 Are there any things in the box that you own? Which adjectives can go with them?

engagement ring, antique, gold

3 Read *I couldn't live without it* and write down the nouns in the box which answer the questions in the passage.

4 There may be some unfamiliar vocabulary in the passage. First decide what part of speech the word is – noun, adjective or verb. Then answer the questions and try to guess the meanings of the words or phrases from the context.

remote control – What does Lesley use it for?

prints – Where does Helen keep them? What can she see in them?

hunt – If you can't find an ashtray but you need one and you know there's one in the house, what do you do?

grab – If you had to leave in a hurry and a favourite possession was near, what would you do to it?

GRAMMAR

> ### Order of adjectives
>
> **When there is more than one adjective before a noun, the order of the adjectives is usually:**
>
> | opinion | size/age/shape | colour | origin | material | purpose | NOUN | other things |
> |---|---|---|---|---|---|---|---|
> | *lovely,* | *old,* | | | *silver* | | *pen* | |
> | | *large,* | *white,* | *French* | | *coffee* | *cups* | |
> | *beautiful,* | *antique,* | | | *gold* | | *ring* | *with rubies* |
> | *nice,* | *rectangular* | | | | | *frames* | |
>
> **When there is a list of adjectives, you usually put a comma after each one except the one before the noun.**
>
> ### Possessive pronouns and adjectives
>
> *my – mine, your – yours, his – his, her – hers, it – its, our – ours, their – theirs*
>
> **We use a possessive pronoun when the noun is understood.**
> *We've got two prints.* **Mine** *is older than my husband's, but* **his** *is worth more.*
> *It was my grandmother's ring, but it's* **mine** *now.*

1 Put the adjectives in the right order.

1 a /wooden/round/charming/dining table
2 a /smart/Italian/white sports car
3 a /paper/large/old/brown bag
4 two /leather/English/comfortable armchairs
5 sparkling/fine/German/wine

2 Work in pairs. Describe to your partner the possessions you made notes on in *Vocabulary and reading* activity 2. Make sure you put the adjectives in the correct order.

3 Choose the correct possessive pronoun or adjective.

1 Give me *your/yours* address and I'll give you *my/mine*.
2 *My/mine* coat is the brown one. Which one is *your/yours*?
3 We've got *our/ours* luggage but they haven't got *their/theirs* yet.
4 It was *her/hers* book but she lent it to *my/mine* husband.
5 Have you got a flat of *your/yours* own? No, I share with *my/mine* friend.

SOUNDS

Listen to these compound nouns. Which word is stressed in compound nouns?

alarm clock   coffee cup
fountain pen   ashtray   hairbrush
picnic basket   wristwatch

Now say the words aloud.

SPEAKING

1 Work in pairs and talk about your own answers to the questions in *I couldn't live without it*. Describe your favourite possessions in as much detail as possible.

2 Look at the objects your teacher has placed in front of you. Try to remember as much detail about them as possible. When your teacher has covered the objects, make a list of them, and describe them in as much detail as possible. Use adjectives in the correct order.

VOCABULARY

1 There are many adjectives and participles which are followed by *to, of* and *with* + object.

She looks very similar to her mother.
It was full of water.
She got bored with him.

Put *to, of* or *with* with these adjectives.

> accustomed afraid allergic angry bored capable
> certain confident fond frightened in love pleased
> proud related similar typical

2 Complete these sentences with an adjective from 1. There may be more than one possible answer.

1 She didn't like swimming because she was ___ of water.
2 Your teacher is very ___ with your hard work.
3 After several years, she grew ___ to his strange behaviour.
4 I am ___ of her support in the election.
5 She was ___ to cats.
6 It was ___ of him to lose his temper.

3 There are some adjectives which usually only go after a noun. For example, you can say:

The man was asleep. not ~~*The asleep man.*~~

In this list are some of the most common adjectives which can usually only go after a noun. Which are they?

> able afraid alive alone apart generous glad kind
> ill likely ready sorry sure unable well

Write the adjectives in sentences which show their meaning.

She feels so ill she's unable to come to work.

4 Look at the vocabulary boxes for lessons 11–15 again. Choose words which are useful to you and write them in your *Wordbank*.

GRAMMAR

1 Complete the sentences with *already, yet, still.*

1 No I haven't found my pen ___ . I'm ___ looking for it.
2 She only started the job a few weeks ago and she's ___ got bored.
3 He hasn't arrived ___ . I expect he's ___ stuck in a traffic jam.
4 I sent the invitation two weeks ago, and I'm ___ waiting for a reply.
5 We're very late. I hope the film hasn't started ___ .
6 Doesn't time fly? It's ___ September and it seems we haven't had any summer weather ___ .

2 Write sentences using the present perfect continuous and *for* or *since.*

1 It started raining an hour ago. It's still raining.
2 I came to this flat twenty years ago. I still live here.
3 I got to the bus stop twenty minutes ago. I'm still waiting for a bus.
4 They started building the house three months ago. They're still building it.
5 She arrived ten minutes ago. She's still standing there.
6 He started painting the room a week ago. He's still painting it.

1 It's been raining for an hour.

3 Rewrite the dialogue with the present perfect simple or present perfect continuous form of the verb in brackets.

TONI: George! How are you? I (not see) you for years!

GEORGE: Toni! Nice to see you! Where (be) you all these years? (Work) you abroad?

TONI: No, I (run) my own business in the West Country. I started it several years ago and it (get) more and more successful in the last few years.

GEORGE: Congratulations! And do you still live in London, or (move) you?

TONI: No, we (move) so that I'm closer to the office.

4 Rewrite the sentences beginning with the words in brackets.

1 Paris is smaller than New York. (New York)
2 In Britain there are fewer tourists than in Spain. (In Spain)
3 In Portugal there is less industry than in Germany. (In Germany)
4 London is not as beautiful as Paris. (Paris is)
5 Madrid is less dangerous than New York. (New York is)
6 There is more rain in Manchester than in London. (In London)

1 New York is bigger than Paris.

5 Think of four things you own and write a description of them using as many adjectives as possible.

We have a lovely, old, fat, black cat.

6 Choose the correct possessive pronoun or adjective.

1 Have you got *my/mine* pen?
2 I don't think it's *her/hers* car. It must be *their/theirs*.
3 We don't have *our/ours* passports. Have you got *your/yours*?
4 I don't recognise *their/theirs* faces. What are *their/theirs* names?
5 Did you get *our/ours* tickets? I'll pay you for *my/mine*.
6 Let's go back to *my/mine* house and catch up on *your/yours* news.

SOUNDS

1 Some words contain two or more consonants and may be difficult to pronounce. Listen and repeat these words.

crisps   knives   clothes   notes   hostages   presidents
terrorists   troops   interests   sports   politics

2 In *Progress check* lessons 6–10, you saw some of the types of sounds which you link in connected speech. Listen to some more types of sounds which you link. As you listen, say the sounds aloud.

– /d/ and /n/ change when they come before /p/, /b/, /w/, /m/, /k/ or /g/

handbag   porcelain bowl   fountain pen   red plastic pen case   gold watch   pinball   good boy

– /t/ disappears when it comes before any consonant

first class   wristwatch   fast lane   waste paper
postbox   most stupid   district cemetery   I must go

3 Read this news story and underline the words you think the speaker will stress.

A demonstration against unemployment has been taking place in Manchester. Demonstrators have been marching through the city for two hours. It is expected to finish in front of the town hall at five this evening where left-wing Members of Parliament will address the crowd.

Listen and check. Read the news story aloud.

SPEAKING

1 Work in groups of three or four. You're going to play *The Drawing Game*. You need a pen and a pad of paper for each group.

The Drawing Game – How to play

1 One person from each group goes to the teacher. Your teacher will whisper a word to you which uses one of the structures presented in lessons 11–15.

2 Return to your group and, without saying or writing the word, do a drawing to illustrate it. The other members of the group have to guess what the word is. You can only answer *yes* or *no*.

3 The person who guesses correctly goes to the teacher, whispers the word, and if the teacher agrees that it's correct, he or she will be given a new word.

4 The first group to guess all the words correctly is the winner.

37

VOCABULARY

1 Work in pairs. Put these words under the following headings: *hair, face, clothes, accessories.* Some words can go under more than one heading.

stubble baseball cap waistcoat belt bikini bow tie flip-flops cardigan earrings flares T-shirt handbag high heels jeans vest crew cut moustache ponytail leather jacket boxer shorts beard sandals sunglasses

2 Look at these words for shades of different colours.

chocolate stone crimson scarlet navy beige olive khaki turquoise canary cream

Match the colours and their shades.

red yellow black blue brown grey green white orange

Now turn to Communication activity 15 on page 102 to check.

3 Which of these materials are fashionable in your country at the moment?

leather cotton linen silk canvas towelling wool denim

Which items in 1 and colours in 2 are fashionable for men or for women at the moment in your country?

Aix is a university town, and there is clearly something that attracts pretty students. The terrace of the Deux Garçons café is always full of them, and it is my theory that they are there for education rather than refreshment. They are taking a degree course in café behaviour, with a syllabus divided into four parts.

One: the arrival You must always arrive as conspicuously as possible, preferably on the back of a crimson Kawasaki 750 motor cycle driven by a young man in head-to-toe black leather and three-day stubble. You mustn't stand on the pavement and wave him goodbye as he drives off down the street to visit his hairdresser. That is for naïve little girls from the Auvergne. The sophisticated student is too busy for sentiment. You are concentrating on the next stage.

Two: the entrance You must always keep your sunglasses on until you see a friend at one of the tables, but you should not appear to be looking for company. Instead, the impression should be that you're heading into the café to make a phone call to your titled Italian admirer, when – what a surprise! – you see a friend. You can then remove the sunglasses and toss your hair while your friends persuade you to sit down.

Three: ritual greetings You must kiss everyone at the table at least twice, often three times, and in special cases four times. Your friends are supposed to remain seated, allowing you to bend and swoop around the table, tossing your hair, and getting in the way of the waiters.

Four: table manners When you have sat down, you should put your sunglasses back on to allow you to look at your own reflection in the café windows – to check important details of technique: the way you light a cigarette, or suck the straw in a Perrier, or nibble daintily on a sugar lump. If these are satisfactory, you can adjust your glasses downwards so that they rest charmingly on the end of the nose, and attention can be given to the other people at the table.

This performance continues from mid-morning until early evening, and never fails to entertain me. I imagine there must be the occasional break for academic work in between these hectic periods of social study, but I have never seen a textbook on any of the café tables, nor heard any discussion of philosophy or political science. The students are totally involved in showing form, and the café terrace is all the more decorative as a result.

Adapted from *A Year in Provence* by Peter Mayle

Do it in style

READING

1 Read *Do it in style* and decide if the writer's advice is mainly for men or for women.

2 Is the style of the passage serious or humorous? Do you think the writer approves or disapproves of the students' behaviour?

3 There may be some unfamiliar vocabulary in the passage. Answer the questions and try to guess the meanings of the words or phrases.

conspicuously – Would someone who was conspicuous be discreet or noticeable?

titled – If you wanted to be cool, would you want a common, ordinary admirer or one who was a noble man or woman?

swoop – Is this movement likely to be one where you keep still or lean over in different directions?

tossing – If you were proud of your long hair, would you shake it over your head or keep it still?

nibble daintily – Is this likely to mean eat carefully or eat greedily?

hectic – Is this likely to mean something like calm or busy?

FUNCTIONS AND GRAMMAR

> Asking for and giving advice; *must, should*
>
> **You use *must* for giving strong advice.**
> *You **must** always arrive as conspicuously as possible.*
> *You **mustn't** stand on the pavement and wave him goodbye.*
>
> **You use *should* or *(be) supposed to* to give less strong advice.**
> *You **should** put your sunglasses back on.*
> *You **shouldn't** appear to be looking for company.*
> *They **are supposed to** remain seated.*
> *You**'re not supposed to** talk about your studies.*
>
> **You can use these expressions to ask for advice:**
> ***Do I have to*/*need to** kiss everyone?*
> ***Should I/Am I supposed to** wear sunglasses inside?*
> **You don't usually use *must* in questions.**
>
> ***Must*** and ***should*** **are modal verbs. Remember that they have the same form for all persons, don't take the auxiliary *do*, and take an infinitive without *to*.**

1 Work in pairs. What advice would you give to someone in your country who wanted to make a stylish impression, about:

1 the arrival   3 ritual greetings
2 the entrance   4 table manners

1 You must arrive in a Porsche.

2 Write sentences giving advice on how to do the following things in style.

1 travel to Venice   4 spend the weekend
2 attract someone's attention   5 go shopping
3 entertain important guests   6 give a party

1 You should take the Orient Express and you must dress in clothes from the thirties.

SPEAKING AND WRITING

1 Work in pairs. What's fashionable in your town or country at the moment? Talk about:

restaurants   bars   music   clothes   fashion accessories entertainment (eg films)

2 Write some advice to visitors to your country about clothing, accessories and behaviour in the following situations.

– in an office   – in church   – at a dinner party
– at weekends   – at school   – on the beach
– in a bar

You're not supposed to wear jeans in an office.

Making predictions; *may* and *might*
going to and *will*

FUNCTIONS AND GRAMMAR

Making predictions; *may* and *might*

Certain

*I'm **certain/sure** he'll be here at nine tomorrow.*
*He's **certain/sure to be** here at nine tomorrow.*

Probable

*She'll **probably** arrive on time.*
*She **probably won't** be late.*

Possible

*I **may/might** be late tomorrow.*
*It's **possible that** I **won't** get there on time*
***Maybe/perhaps** I'll call and see some friends.*

May is a little more sure than ***might***. ***May***,
might and ***will*** are modal auxiliary verbs.
(See Lesson 16.)

Going to and *will*

**You use *(to be) going to* to talk about things
which are arranged or sure to happen, and for
decisions which you made before the moment
of speaking.**
*I'**m going to** spend Christmas in Tenerife.*

**You use *will* to talk about decisions you make
at the moment of speaking.**
*'We'**ll stop** for today,' he said.*

1 Choose the correct verb form.

1 I *won't/might not* be in at nine tomorrow as I
have a dentist's appointment at that time.

2 There are no buses after midnight, and it's too
far to walk so I *may/will* take a taxi.

3 She *is going to/might* come and stay at
Christmas or she may stay at home.

4 I'm thinking of seeing some friends tonight,
so I definitely *won't/may not* be back at seven.

5 They*'re going to/'ll* go to the theatre tonight.
They've got tickets.

2 Think of three things which will certainly
happen tomorrow, three things which will
probably happen, and three things which will
possibly happen. Use expressions from the box
above.

3 Write down three decisions about things you'll
do tomorrow. Use *will*.

I'll buy an English newspaper.

Now tell other students about your decisions.
Use *going to*.

I'm going to buy an English newspaper.

LISTENING AND SPEAKING

1 Work in pairs. You're going to listen to and read a story in three parts called *Rose Rose*, by Barry Pain. First, look at the picture. When and where might it take place? What might it be about?

It might be about an artist.

What type of story might it be?

– a love story   – a science fiction story
– a ghost story   – a detective story

2 In part 1 of the story there are two characters: Sefton, the artist and Miss Rose, his model. Read the sentences below. Who do you think is speaking?

a   'Tomorrow, at nine?'
b   'Rest now, please,'
c   'I'll be here by a quarter past nine. I'll be here if I die first!'
d   'You think I don't mean to come tomorrow.'
e   'Oh, you don't want any money. You're known to be rich.'
f   'The trouble with these studios is the draughts.'
g   'We'll stop for today,'

Can you imagine what might happen in the first part of the story?

3 🔊 Listen to part 1 of the story. As you listen, put the sentences in 2 in the right order. Did you guess correctly?

4 Work in pairs. Here are some questions about part 2 of the story. Imagine what the answers might be.

1   What time does he get to his studio next morning?
2   Why does Aphrodite have a mocking expression?
3   Where does he go when he finds he has no tobacco?
4   What time does he expect Miss Rose to arrive?
5   When does she arrive?
6   When he suggests a rest, what does he feel?
7   Where is Miss Rose when he looks round?
8   How does he feel now?
9   What does he read in the newspaper?
10   What is he going to do with his pocket knife?

5 🔊 Listen to part 2 of the story and check your answers to 4.

VOCABULARY AND WRITING

1 Here are some words from the story. Check you know what they mean.

> model   stove   brush
> draught   depend upon
> beautiful   screen   sitting
> picture   mocking   tobacco
> studio   flat   fatigue
> disappear   knock down
> pocket-knife

2 Work in pairs. How do you think the story ends? What do you think might happen to Sefton? Does he see Miss Rose again? Write a suitable ending to the story. Think about:

– what Sefton might do to the painting
– where Miss Rose might be
– what might happen when he starts painting again
– what might happen to him

You may want to contrast some ideas. You can use linking words *although* and *in spite of* + noun or *-ing* when the subject of the two clauses is the same.

Although she was dead, she returned the next day.
In spite of being dead, she returned the next day.

When the subject of the two clauses is different you can only use *although*.

Although she was dead, he felt her presence in the studio.

3 To find out how the writer finished the story, turn to Communication activity 8 on page 99. Do you think your ending is better?

18 | *What do you do for a living?*

Drawing conclusions; *must, can't, might, could*; describing impressions

VOCABULARY AND SPEAKING

1 Look at these words for jobs.

> teacher nurse vet journalist designer hairdresser secretary bank worker
> receptionist police officer soldier sailor farmer gardener park keeper
> waitress photographer

What do you need to be to do the jobs? Choose from the words below.

> imaginative sociable hard-working intelligent reliable funny lively kind
> fun fashionable flamboyant sensible sensitive smart practical young
> organised fit

You need to be intelligent to be a teacher.

In your job or ideal job...

Do you work...?

inside ☐	outside ☐
in an office ☐	at home ☐
on your own ☐	at weekends ☐
in the evenings ☐	at nights ☐

Do you need to be...?

imaginative ☐	ambitious ☐
attractive ☐	young ☐
physically fit ☐	

Do you...?

travel ☐
need to drive ☐
wear a uniform ☐
work with your hands ☐
use a tool ☐
use a machine ☐
make or build anything ☐
need to type ☐

Do you have to...?

talk a lot ☐
meet a lot of people ☐
have special qualifications ☐
speak a foreign language ☐
give orders or instructions to others ☐

2 Read and answer the questionnaire.

3 Work in pairs and talk about your answers to the questionnaire.

FUNCTIONS

Drawing conclusions

Certain

***Must* + infinitive**
He's worked with her for years.
*He **must know** her very well.*

***Can't* + infinitive**
He never sees anyone.
*He **can't like** people.*

Possible

***Might* + infinitive**
*He **might be** an accountant.*

***Could* + infinitive**
*She **could be** Italian.*
*She **could work** in sales.*

In this context *might* and *could* mean the same.

Describing impressions

***Look/sound* + adjective**
*She **looks/sounds** nice.*

***Look/sound like* + noun**
*He **looks/sounds like** a nice person.*

***Look/sound as if* + verb**
*She **looks/sounds as if** she works very hard.*

42

1 Agree with the conclusions using *must*, *can't* and *might*.

1 I'm sure he's a journalist.
2 He certainly isn't a doctor.
3 Perhaps he's working late.
4 I'm certain she doesn't have a job.
5 I'm sure she's able to drive.
6 I'm certain she isn't in the office at the moment.
7 He's possibly a farmer.
8 I'm certain she's reliable.

1 Yes, he must be a journalist.

2 Write sentences using *look/sound* or *look/sound like* or *look/sound as If* and the word in brackets.

1 You've got red eyes. (tired)
2 She's got a very quiet voice. (shy)
3 He's got big feet. (police officer)
4 I can hear something outside the front door. (cat)
5 Our neighbours are very noisy. (dance)
6 He's reading a book. (work)

1 You look tired.

SOUNDS

1 Say these words aloud. Underline the stressed syllable.

journalist designer secretary receptionist
police officer soldier gardener

🔲 Now listen and check.

2 🔲 Listen to the way Speaker B uses a strong stress to disagree.

1 A He can't be a journalist.
  B He must be a journalist.
2 A He must be a doctor.
  B He can't be a doctor.
3 A He can't be working late.
  B He must be working late.
4 A She must have a job.
  B She can't have a job.

Now work in pairs and say the sentences aloud.

LISTENING

🔲 Listen to three people answering the questionnaire. Your teacher will stop the tape after every two or three questions. Try and guess what each person's job is. Choose from the jobs in *Vocabulary and speaking* activity 1.

READING

1 The passage below consists of two texts in which two people talk about their jobs. The texts have been mixed up. First, read the passage and decide what the two jobs are. Choose from the jobs in *Vocabulary and speaking* activity 1.

Now work in pairs and check your answers.

2 Read the passage again and separate the two texts by underlining the sentences which belong to the same text. The sentences are in the right order.

It can be very tiring and the hours are very long because some people don't leave until two or three in the morning. People think it's an exciting job, but it's not really, just hard work. I usually get to the studio at about ten in the morning, and open the mail, make a few phone calls, plan my work for the next few months. Then you have to get some sleep and be ready to start again at about half past eleven the next morning. Then my assistant arrives and he starts getting the equipment ready for the session which usually begins at about eleven o'clock. Some of the customers can be really difficult, complaining about everything, but usually they're only showing off to their friends. The most interesting ones are the fashion shoots, but the ones I do most often are what are known as pack shots, where I take photos of objects. For example, it could be an advertising campaign for a brand of soap powder or maybe a new car. For example, we had one man here who sent back three bottles of wine, saying that they were no good. The manager had to tell him the wine was fine, which the man didn't like at all. Two or three times a year I travel abroad, and do some location work, which I enjoy. We don't get paid very much and people can be very mean with their tips. But on a good night it can be good fun.

Now work in pairs and check your answers.

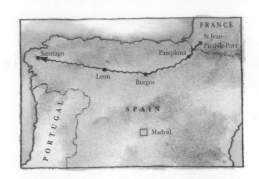

The Way of St James

IF YOU DRIVE across the south-west of France and into northern Spain you'll begin to notice groups of walkers. Most are carrying backpacks and long sticks, and somewhere they're wearing a scallop shell. You can stop and talk to them, but you're not allowed to give them a lift. They're on a guided tour, but they're not ordinary tourists. They're pilgrims walking the *Camino de Santiago*, the Way of St James, to Santiago de Compostela, the route that millions of people have taken over hundreds of years.

The pilgrimage to Santiago, where St James is believed to be buried, was extremely popular, especially among the French and other north Europeans because it was easier to get to than the pilgrim routes to Rome or Jerusalem.

In the Middle Ages, pilgrims were rather like today's package tourists. The season began in April or May, and they travelled in groups because it was safer and also more enjoyable. To prove they had done the pilgrimage, the rules were quite strict. They had to follow a well-planned route and visit important places of culture where they bought souvenirs – scallop shells, for example – to prove where they had been. They had to travel on foot or by horse and they stayed in special hostels. For some, the pilgrimage was an important religious experience, but for many it was a chance to have a holiday and do some sightseeing.

These days the rules are less strict. You only have to travel 100 km on foot or horseback. You can go by bicycle as well, but you're not supposed to drive or hitchhike, so to prove you have resisted this temptation, you're obliged to obtain a 'passport' from the Confraternity of St James and get a stamp at various offices along the route.

You don't even have to be very religious. Many people see it as an alternative to package tours and beach holidays. But the *Camino* is getting very popular and if you want to do it in the peace and quiet which the pilgrims in the past enjoyed, you should travel out of season and avoid the fiesta of St James in July.

When you arrive in Santiago, you have to show your passport at the Pilgrim's Office by the cathedral. A church official checks the dates on the stamps and if he is satisfied, he gives you a *compostela*. The *Hotel de los Reyes Catolicos,* where the pilgrims used to stay and which is now a modern hotel, is still obliged to give up to ten pilgrims one free meal a day for three days, although they can't eat it in its main, very fashionable restaurant. Finally, you're supposed to enter the magnificent cathedral and touch the statue of St James. With this last gesture you have become part of the pilgrim tradition that has attracted believers and tourists for many hundreds of years.

Adapted from *A Pilgrim's Package* by David Lodge,
The Independent

VOCABULARY AND READING

1 Look at these words and put them under two headings: *religion* and *tourism.*

backpack bicycle cathedral church hitchhike
holiday horseback hostel incense package tour
pilgrim priest saint shell sightseeing souvenirs
stamp statue temptation walker

2 Read *The Way of St James* and choose the best definition.

The Way of St James is:
a a holy path around the centre of Santiago
b the route of the pilgrims to Santiago
c a package tour around northern Spain
d a religious festival

3 Which statement describes the link the writer makes between religion and tourism?

1 People went on a pilgrimage in the past in the same way people go on package tours today.
2 Religion today is like mass tourism.
3 Most tourists today look like pilgrims from the past.

FUNCTIONS

Talking about obligation
You **have to** *show your passport.*
You**'re obliged to** *obtain a passport.*
Have to is less formal than be obliged to.
You can use *be supposed to* to talk about what people are expected to do because of a rule or a tradition.
You **are supposed to** *enter the magnificent cathedral.*
Talking about absence of obligation
You **don't have to** *be religious.*

Talking about permission
They**'re allowed to** *buy their shells when they start their journey.* You **can** *go by bike as well.*

Talking about prohibition
They **can't** *eat the meal in the main restaurant.*
You**'re not allowed to** *give them a lift.*
You can use *(be) not supposed to* to suggest a difference between what is not allowed and what may happen.
Pilgrims **are not supposed to** *drive or hitchhike.*
(It's not allowed, but some may do so.)

1 Correct these false statements about *The Way of St James*. Write full sentences.

1 You don't have to give the pilgrims a lift.
2 You're not allowed to talk to them.
3 They didn't have to travel in groups.
4 They weren't allowed to stay in special hostels.
5 They were supposed to travel on foot.
6 You're supposed to get a stamp at various places.
7 You're allowed to show your passport in Santiago.
8 Pilgrims have to eat in the main hotel restaurant.

Read the passage again and check your answers.

2 Rewrite these rules with expressions from the functions box.

1 No smoking.
2 Please show your tickets.
3 No photography.
4 Women: cover your head.
5 Talk quietly.
6 Please remove your shoes.

SOUNDS

The pronunciation of *can, have to* and *must* is weak when a verb follows and strong when they are on their own. Listen and repeat these sentences. Make sure you use a weak or strong pronunciation.

1 You have to show your passport. 2 Do I have to?
3 They can walk or ride. 4 Yes, they can.

WRITING

1 Think of a situation which involves a number of rules in your country. Make notes on:

– what you have to do
– what you're allowed to do
– what you aren't allowed to do
– what you aren't supposed to do

2 Write a description of the situation you chose in 1. Use the linking words in **bold** for a list of points.

When you do your military service, you're obliged to join the army for a year. **Firstly**, *you're obliged to stop your studies.* **Then** *you have to leave your family.* **What's more**, *you aren't supposed to go home very often. And* **worst of all**, *you aren't allowed to have long hair.*

Use *however* to introduce contrasting points. Remember to separate it from the rest of the sentence with a comma.
However, *you're allowed to go out at weekends.*

20 | *How unfair can you get?*

Talking about ability and possibility; *can, could, be able to*

VOCABULARY AND SPEAKING

1 Answer these questions about personal qualities.

Can you ...

always tell people what you really think? ☐
relax with people you don't know? ☐
usually get what you want? ☐
keep calm in stressful situations? ☐
keep your temper under control? ☐
laugh at yourself? ☐
always see both sides of an argument? ☐
ignore criticism easily? ☐
express your feelings easily? ☐

2 Work in pairs. Tell each other what you can or can't do.
Are you similar or different?

I can't always tell people what I think. Can you?   No, I can't.

3 Match the descriptions in 1 with compound adjectives from the list below. There may be more than one possibility.

> outspoken   short-tempered   middle-aged   good-humoured   thick-skinned
> fair-minded   easy-going   outgoing   cool-headed   strong-willed
> soft-hearted   self-assured   well-behaved   hard-working   old-fashioned
> short-sighted   left-handed   world-famous

Write sentences that describe each compound adjective.

Someone who can always tell people what they really think is outspoken.
Someone who can't keep their temper under control is short-tempered.

How would you define the other adjectives in the box?

Someone who is middle-aged is someone who is neither young nor old.

Look at the people in the pictures and choose compound adjectives to describe them.

4 Have you always had the qualities you talked about in 1 and 2?
Did you have them when you were a child?

When I was ten I couldn't laugh at myself. I was very serious.

You can use these adjectives to help you.

> unemotional   open   deceitful   tactful   reserved   honest   direct   assertive
> aggressive   fair   shy   confident   cool   sensitive   serious   prejudiced

FUNCTIONS AND GRAMMAR

> Talking about ability and possibility: *can, could, be able to*
>
> *Can* and *could* are modal verbs. (See Lesson 16).
> You use *can* to talk about general ability in the present.
> *I can usually get what I want.*
> *I can't always say what I think.*
>
> You use *could* to talk about general ability in the past.
> *When I was young I could usually get what I wanted.*
> *I couldn't laugh at myself.*
>
> You use *be able to* to talk about ability in the other tenses.
> *Has he been able to calm down?*
> *She'll be able to relax with them, won't she?*
>
> You use *was/were able to* when something was possible on a specific occasion in the past.
> Here it means *managed to do something.*
> *He was angry but he was able to keep his temper.*
> *The weather was good so we were able to go out.*
>
> But you can use *couldn't* for general ability and specific occasions.
> *He was so angry that he wasn't able to/couldn't keep his temper.*
>
> You use *can* and *could* with verbs which don't usually go in the continuous form.
> *It was raining so we weren't able to/couldn't go out.*

1 Complete the sentences with the correct form of *can, could* or *be able to*. Sometimes more than one answer is possible.

1 I know your face but I (not) ___ remember your name.
2 He looked for the book but he (not) ___ find it.
3 When he was young, he ___ sing very well.
4 She's ill, so she won't ___ come this weekend.
5 If he spoke slowly, I ___ understand him.
6 It was very dangerous, but she ___ keep calm.
7 She was sensitive and she (not) ___ take criticism.
8 I used to ___ dance well, but I (not) ___ dance last night.

2 Think of two or three specific occasions in the past when you were or weren't able to do something you wanted or needed to do. Say what happened.

Usually I can't relax with people I don't know well, but when I met Philip, he was so friendly that I was able to get on very well with him immediately.

LISTENING AND WRITING

1 You're going to hear Rosalind, an English woman, describing an occasion in the past when she lost her temper. First, look at these words and phrases from part 1 of her story. What do you think happened?

friend   lunch   parking space   multi-storey car park   lift
loud noise   stuck   no lights   eight people   alarm bell
emergency light   whimper   panic   poke in the eye

🔘 Now listen to part 1 and check your answers to 1.

2 Are these statements true or false? Correct the false ones.

1 Rosalind couldn't drive. ☐
2 They couldn't find anywhere to park. ☐
3 They took the stairs. ☐
4 Peter couldn't walk. ☐
5 They couldn't open the doors. ☐
6 They couldn't hear the alarm bell. ☐
7 They thought of lots to talk about. ☐
8 She couldn't find the control panel. ☐
9 She was able to control her temper. ☐
10 She couldn't laugh about it afterwards. ☐

🔘 Listen to part 1 again and check.

3 Work in pairs. What do you think happened next?

4 🔘 Listen to part 2 of the story and put a tick (✓) by the things the people were able to do and a cross (✗) by the things they weren't able to do.

get back to the car in time ☐   join in the argument ☐
persuade him to take the ticket back ☐   pay the fine ☐
keep her temper ☐   disappear into the crowd ☐
take the ticket from the windscreen ☐   control herself ☐

5 Work in pairs and check your answers to 4.

🔘 Listen again and check. Try to remember as many details as possible.

6 On a piece of paper, start to write the story of what happened to Rosalind. Use the words in 1, the statements in 2 and the points in 4 to help you.

When you have written a few sentences, pass your piece of paper to another student. You will receive a piece of paper with sentences from someone else.

Read the sentences, correct any mistakes, then continue to write the story. Continue giving and receiving pieces of paper like this, until you finish the story.

Progress check 16–20

VOCABULARY

1 Multi-part verbs, or phrasal verbs, consist of a verb followed by one or two particles. They are very common in English. There are four types of multi-part verbs.

Type 1 verbs take no object.
look out, sit down

Type 2 verbs take an object.
turn down, take off
The noun object can go either before or after the particle.
*Can you turn the radio **down**?*
or *Can you turn **down** the radio?*

But the pronoun object must go between the verb and the particle.
*Can you turn **it** down?*

Type 3 verbs take an object.
look after, join in

The object must go after the particle.
*I'll look after **the children**.*
*I'll look after **them**.*

Type 4 verbs have two particles.
get on with, go down with

The object must go after the particle.
*I get on with **Jean**.*
*I get on with **her**.*

Here are some multi-part verbs. Use a dictionary to find out what they mean and decide if they can take an object or not; if so, decide where the pronoun object goes. Some may be more than one type.

take off   bring up
give up   pull down   pick up
give back   fill in   turn down
put off

2 Multi-part verbs are more common in English than verbs with a similar meaning. But if you find the multi-part verb difficult, people will usually understand you if you use a verb with a similar meaning. Match the multi-part verbs in 1 with these verbs which have the same meaning.

demolish   complete (a form)   lift   postpone   raise (children)
reduce (sound, light)   remove   return   stop (doing something)

Write three sentences using the verbs with noun objects and three sentences with pronoun objects.

3 Complete these sentences with suitable multi-part verbs from the list. Make sure you put the noun or pronoun in the right position, and the verb in the right tense.

bring back   clear up   call off   cut out   give away   go over   join in   open out

1 When you've read my book, could you ___ it, please?
2 She ___ the insurance document very carefully.
3 They asked if he wanted to ___ the football game.
4 The kitchen was in a mess so I ___ it.
5 The article was too long, so he ___ two paragraphs.
6 If I had a lot of money I would ___ it all.

Can you rewrite the sentences using verbs with similar meanings?

Now write sentences using the two extra verbs.

4 Look at the vocabulary boxes for lessons 16–20 again. Choose words which are useful to you and write them in your *Wordbank*.

GRAMMAR

1 Choose the correct verb form.

1 'Would you like a drink?' 'Thank you. I'll have/I'm going to have some juice.'
2 Look at those dark clouds. It'll/It's going to rain.
3 We'll/We're going to live in Paris. We've just bought a flat there.
4 I feel tired. Perhaps I'll/I'm going to sit down.
5 It's John's birthday, so I'll/I'm going to buy him a present.

2 Agree with the sentences using *might, must* and *can't.*

1 I'm sure she's tired.
2 Perhaps it's late.
3 Maybe he's stuck in the traffic.
4 I'm sure she lives here.
5 Perhaps he doesn't speak French.
6 I'm sure she isn't home yet.

1 Yes, she must be tired.

3 Write sentences using *look/sound* or *look/sound like* and the word in brackets.

1 That's a lovely armchair! (comfortable)
2 He's carrying a lot in his backpack. (heavy)
3 He's got a camera and sunglasses. (tourist)
4 Can you hear the bells? (church)
5 He's always making jokes. (good-humoured)
6 He gets angry very quickly. (short-tempered)

1 *That armchair looks comfortable.*

4 Choose the correct form of the verb.

1 You *can't/don't have to* park on a pedestrian crossing.
2 You *shouldn't/can't* stay up so late if you're so tired in the morning.
3 'Keep off the grass' means that you *can't/don't have to* walk on the grass.
4 You *mustn't/don't have to* be late for work.
5 Thank you for the lift, but I'm afraid you *can't/don't have to* come in.
6 You *mustn't/might not* put your feet on the table.

5 Complete these sentences with the correct form of *can, could* or *be able to*.

1 It was late but luckily we ___ get the last bus home.
2 If she wrote clearly, I ___ read her handwriting.
3 Sorry, ___ you repeat that? I (not) ___ hear what you said.
4 He worked very hard and he ___ earn a lot of money.
5 He was tired, but he ___ stay awake until he got home.
6 It was raining so we (not) ___ go out.

SOUNDS

1 How do you pronounce the following words? Put them in five groups according to the way you pronounce the letter *a*: /æ/, /ɔː/, /ɑː/, /eɪ/ or /eə/.

small   dark   hand   take   father   care   rare   awful   flat   name   sale   call   start   man   stare

[cassette] Listen and check. Say the words aloud.

2 [cassette] Listen and repeat these words.

Spanish   school   stranger   speed   spelling   sport   skirt   spend   smile   stand   spill

Can you think of other words which begin with *s* + consonant?

3 Circle the /ə/ sound. Then underline the words you think the speaker will stress.

If you drive across the south-west of France and into northern Spain you'll begin to notice groups of walkers. Most are carrying backpacks and long sticks, and somewhere they're wearing a scallop shell. You can stop and talk to them, but you're not allowed to give them a lift.

[cassette] Listen and check. Say the sentences aloud.

SPEAKING AND WRITING

1 Read these statements and decide which ones must be true, might be true, and can't be true.

● You don't have to separate clear bottles and green bottles for recycling. They always mix them all up again at the recycling depot.
● Manufacturers are able to make cars which are impossible to break into or steal, but they won't do it because sales of new cars will drop.
● Computers never forget anything that you lose accidentally, it's just that you have to be an expert to get at all the information.
● The most expensive computers know when the guarantee runs out.
● Putting bigger wheels on the back of your car means you're always going downhill, and you're always saving petrol.
● You can recharge phone cards by putting them in the freezer overnight.
● The journey from Paris to Lyon is quicker than Lyon to Paris as it's downhill.

Now work in pairs and check your answers.

2 Work in pairs. Write one statement that must be true, one statement that might be true and one statement that can't be true.

3 Show your statements to the rest of the class. Read the statements written by other people in the class and decide which statements are true and false.

21 Cinema classics

Adverbs (2): formation; giving opinions; emphasising

VOCABULARY AND LISTENING

1 Look at these words for types of films. Think of an example of a film for each type.

> action film western thriller
> science fiction film horror film
> comedy musical love story

action film: Robocop

2 Look at the adjectives in the box. Which ones can you use to describe something you like? Which ones can you use to describe something you dislike?

> amazing appalling simple
> remarkable delightful charming
> impressive awkward far-fetched
> clumsy sensitive gripping horrible
> funny powerful terrible emotional
> fantastic slow extraordinary
> spectacular

3 Choose an adjective from the box in 2 which goes with the films you thought of in 1.

Robocop is a brilliant film.

4  Listen to John talking to Amanda about a film classic, *Casablanca*. Which adjectives does he use to talk about the following?

– the film   – the acting
– the plot   – the ending

5 What was his favourite scene? Why?

Listen again and check.

GRAMMAR AND FUNCTIONS

Formation of adverbs

You form adverbs by adding -ly to most adjectives.
amazing – amazingly   beautiful – beautifully

With adjectives ending in -y you drop the -y and add -ily.
extraordinary – extraordinarily   clumsy – clumsily

With adjectives ending in -le you drop the -e and add -y.
remarkable – remarkably   terrible – terribly

Giving opinions

I thought it was really good.   I found it uninteresting.
I've never seen such a good film.   It's well worth/not worth seeing.
As far as I'm concerned, it's the best film I've ever seen.

Emphasising

You can also use the following adverbs of degree before an adjective to emphasise something.
absolutely amazingly extremely especially extraordinarily particularly really
It's a really exciting film.

1 Make adverbs from these adjectives.

simple   brilliant   remarkable   delightful
charming   impressive   powerful   happy
spectacular

2 Choose the correct adjective or adverb.

1 He thought it was a *spectacular/spectacularly* film.
2 It was *extreme/extremely* exciting.
3 The characters acted rather *clumsy/clumsily*.
4 The plot was *particular/particularly* far-fetched.
5 It's got a *fantastic/fantastically* ending.
6 It's one of the most *remarkable/remarkably* films I've ever seen.

SOUNDS

 Listen to the way the speaker emphasises an opinion.

1 It's really amazing!
2 It's an absolutely terrible film!
3 The ending is especially exciting!
4 The plot is particularly well-written.
5 The acting is simply appalling.
6 The music is extremely good.

Now say the sentences aloud. Use stress to emphasise your opinions.

WRITING AND SPEAKING

1 Read the plot of *Casablanca* and decide where these adjectives and adverbs can go. (Many can go in more than one position.)

brave   deeply   exotic   extremely   finally   gripping
passionate   precious   reluctantly

> Casablanca is a thriller and a love story with Humphrey Bogart and Ingrid Bergman. The film takes place during the Second World War in Casablanca, a city in Morocco. Rick Blaine, played by Humphrey Bogart, owns *Rick's café* which is a centre for war refugees who are waiting for visas to escape to America. Rick discovers that his former love, Ilsa Lund, played by Ingrid Bergman, is now married to a Resistance worker, Victor Laszlo, whom he is helping to escape. With the enemy on their trail, Ilsa comes to Casablanca to collect the visas which will allow Laszlo to escape and continue his fight for freedom. To her surprise, she finds Rick there. Rick is still in love with Ilsa, but he decides to help the couple escape. It's full of romance, intrigue, and suspense and it's well-filmed. My favourite scene is at the end of the film, at the airport where Rick and Ilsa have to say goodbye and she leaves with her husband on a plane just before the enemy arrives.

2 Work in pairs. Talk about a cinema classic you know well and like. Explain:

– what it's called
– who's in it
– where it takes place
– what your favourite scene is
– what type of film it is
– who it's by
– what it's about
– why you like it

Make notes about your partner's answers.

3 Write a description of your partner's cinema classic. Use the passage in 1 and your notes in 2 to help you.

Wild and beautiful

Adverbs (3): position of adverbs and adverbial phrases

I saw my first tiger in a national park in India. It was a young male and he was drinking at a waterhole. He raised his head slowly and stared at us for a full minute. Then he turned his back on us and disappeared quickly into the jungle.

Twenty years ago, the tiger was in trouble. In India its numbers were around 1,800. Then the Indian government launched *Project Tiger*, which set up national parks all over the country. Poachers still hunt the tiger illegally, but at least it is no longer in danger of extinction.

In Africa, the most important species in danger is the elephant, the world's largest living land mammal. In 1979 there were 1.3 million elephants there. Ten years later, numbers were down to fewer than 600,000 and still falling. Conservationists warned that the species could be extinct by the end of the century.

But slowly the situation changed. In July 1989, Kenya's President Moi publicly burnt his country's stock of ivory, and towards the end of 1989 the world agreed to ban the ivory trade completely. Since then, the demand for ivory has fallen sharply, and elephant numbers in countries such as Kenya and Tanzania are increasing rapidly. In Kenya's vast Tsavo national park, breeding herds of elephants are a common sight, the new babies hurrying to keep up with their mothers.

Africa has lost 99 per cent of its black rhinos in the past twenty years. Ten years ago, there were only 11 rhinos left in Kenya's Masai Mara national reserve. Today, numbers have tripled.

On the other side of the world, the grey whales of Baja California nearly disappeared in the last century. Fortunately, the US Marine Mammals Protection Act of 1972 saved them. That same year, Mexico created the world's first whale sanctuary on the west coast of the Baja. The grey whales recovered quickly. Today there are perhaps 20,000 and these gentle giants are now worth far more alive than dead. The reason is whale-watching, an American craze for tourists.

All over the world other rare species continue to receive protection; giant tortoises in the Galapagos, pink pigeons in Mauritius. In America you can hear the song of the timber wolf, and see the mountain lion in the canyons and high forests.

Suddenly, wildlife is good for the tourist trade. And tourism – provided it takes only pictures and leaves only footprints – is good for the national parks. If wildlife can be seen to be paying its way, then its chance of survival will be much greater.

First printed in British Airways *High Life* magazine

VOCABULARY AND READING

1 In which parts of the world can you find the following animals?

> elephant tiger zebra falcon wolf moose lion giraffe
> whale eagle bear vulture panda leopard gorilla
> rhinoceros tortoise pigeon butterfly jaguar

Which animals are in danger of becoming extinct?

2 *Wild and beautiful* is about animals which may be in danger of extinction. Do you think the description of their situation will be optimistic or pessimistic?

3 Read *Wild and beautiful* and find out which animals it mentions. Are they still in danger of extinction?

4 Are these statements true, false or is there no information in the passage?

1 The Indian government helped save the tiger.
2 The ivory trade put the elephant in danger.
3 The number of black rhinos in the Masai Mara has fallen by a third.
4 Tourists were not interested in whales thirty years ago.
5 There are wild lions in America.
6 Tourism may help save wildlife.

GRAMMAR

> **Position of adverbs and adverbial phrases**
> **An adverb or adverb phrase of manner describes *how* something happens.**
> *He raised his head **slowly**.*
> **An adverb or adverb phrase of place describes *where* something happens.**
> *He was drinking **at a waterhole**.*
> **An adverb or adverb phrase of time describes *when* something happens.**
> ***Then** he turned his back on us.*
> **Adverbs and adverb phrases usually go after the direct object. If there is no direct object, they go after the verb.**
> *He disappeared **quickly**.*
> **If there is more than one adverb or adverbial phrase, the usual order is manner, place, time.**
> *We had set out **by jeep after supper**. (manner, time)*
> *...which set up National Parks **all over the country in a few years**. (place, time)*
> **But you can put some adverbs and adverb phrases before the verb clause for emphasis.**
> ***Twenty years ago**, the tiger was in trouble.*

1 Look at the words and phrases in **bold**. Are they adverb phrases of manner, place or time?

1 **Twenty years ago**, the tiger was in trouble.
2 Poachers still hunt the tiger **illegally**.
3 **In Africa** the most important species in danger is the elephant.
4 **In 1979** there were 1.3 million elephants **there**.
5 The world agreed to ban the ivory trade **completely**.
6 Elephant numbers in Kenya and Tanzania are increasing **rapidly**.

2 Look at these sentences from the passage. Decide where the adverbs and adverb phrases in brackets can go in the sentences.

1 Africa has lost 99 per cent of its black rhinos. (in the past twenty years)
2 There were only 11 rhinos left. (in Kenya's Masai Mara national reserve/ten years ago)
3 The grey whales recovered. (quickly)
4 There are perhaps 20,000. (today)
5 Other rare species continue to receive protection. (all over the world)
6 Wildlife is good for the tourist trade. (suddenly)

Now look back at the passage and check.

SPEAKING AND WRITING

1 Work in groups of three or four. What has changed in your country or in the world as a whole, either for better or for worse? Think about:

– traffic   – green space   – tourism   – pollution
– clean beaches   – motorways   – the climate

2 Write a description of the situation and how it has changed. Use the linking words in **bold**.
Say what the situation was like in the past.
***Ten years ago** we lived in the country, and there were fields at the back of our house.*
Say what changed and why.
***Then** they built a factory, and they needed houses for the workers.*
Say what the situation is like today.
***Today**, although the shops and other facilities are better, there is a lot more traffic.*
Say what the situation will be in the future.
***In ten years' time**, we'll just be part of the city.*
Use adverbs and adverbial phrases where you can.

23 | *Valentine*

Reported speech (1): statements

VOCABULARY AND READING

1 Which of the following are symbols of love in your country?

> apple   rose   onion   moon   heart   spanner   ring
> strawberry   chrysanthemum   snake

Can you think of any others?

2 You're going to read a poem called *Valentine*, by Carol Ann Duffy. In Britain, people often send Valentine cards on St Valentine's Day (February 14th) to someone they love, but they don't reveal their name. Do you have this custom in your country?

3 Read *Valentine* and decide which symbol of love the poet is offering her lover. Choose from the words in the vocabulary box.

4 Answer the questions about some words which may be unfamiliar to you.

blind – If something *blinds you with tears*, are you likely to be unable to hear or unable to see?

wobbling – If you are crying a lot, would your reflection in a mirror be still, or would it appear to shake?

grief – If you are crying, are you usually happy or sad?

cute – Is this likely to mean sentimental and tasteless or attractive?

loop – Is a loop like a wedding-ring likely to be a round shape or square?

cling – Is the scent of an onion likely to disappear quickly or to remain for a long time?

5 Work in pairs. Do you think the poet would agree with these statements?

1 Love is a wonderful, romantic thing.
2 Love will make you cry.
3 Love lasts for ever.

Valentine

Not a red rose or a satin heart.

I give you an onion.
It is a moon wrapped in brown paper.
It promises light
like the careful undressing of love.

Here.
It will blind you with tears
like a lover.
It will make your reflection
a wobbling photo of grief.

I am trying to be truthful.

Not a cute card or a kissogram.

I give you an onion.
Its fierce kiss will stay on your lips,
possessive and faithful
as we are,
for as long as we are.

Take it.
Its platinum loops shrink to a wedding-ring,
if you like.
Lethal.
Its scent will cling to your fingers,
cling to your knife.

Carol Ann Duffy

6 Find words in the poem to go under these headings: *romantic, unromantic.*

romantic: red rose,…
unromantic: onion,…

7 Which of these words would you use to describe the poet's attitude towards her lover?

> honest   direct   aggressive   cruel   possessive   faithful
> jealous   angry   cynical   sarcastic   passionate   unfaithful

How do you think her lover responds to her gift?
How would you respond to it?

LISTENING

 Listen to two people talking about the poem. Which statements in *Vocabulary and reading* activity 5 do they think the poet would agree with?

GRAMMAR

Reported speech (1): statements

You report what people said by using *said (that)* + clause. The tense of the verb in the direct statement usually 'moves back' in the reported statement. Time references also change.

Direct statement	**Reported statement**
'I like him,' she said.	*She said she liked him.*
'I'm seeing him this evening,' she said.	*She said she was seeing him that evening.*
'I've never felt like this before,' she said.	*She said she had never felt like this before.*
'I enjoyed meeting you tonight,' he said.	*He said he had enjoyed meeting her that night.*
'I'll ring you tomorrow,' he said.	*He said he would ring her the next day.*

However, in spoken language, it's common not to change tenses in the reported statement, especially if the statement is still true at the time of reporting.

'I'll always love you,' she said.	*She said she will always love him.*

Time reference changes

today – that day, tonight – that night, tomorrow – the next day, yesterday – the day before, ago – before, last week – the week before, next week – the following week, this morning – that morning

1 Look at the statements from the grammar box. What tense are the verbs in the direct statements and the reported statements?

'I like him,' she said. (present simple)   *She said she liked him.* (past simple)

2 Look at these statements in reported speech. Then look back at *Valentine* and write down what the poet actually said.

1 She said the onion promised light.
2 She said it would blind him with tears.
3 She said she was trying to be truthful.
4 She said its fierce kiss would stay on his lips.

3 Rewrite these sentences in reported speech.

1 'I'm going to call him this afternoon,' she said.
2 'I'll write to you next week,' he said.
3 'He rang me from the airport yesterday,' she said.
4 'I won't be home tonight,' he said.
5 'He left ten minutes ago,' she said.
6 'I'll see him tomorrow,' she said.

4 Write full answers to the *Listening* activity.

SOUNDS

1 How do you think the poem should sound if it is read aloud? Choose from these possible styles.

*loud – soft   fast – slow
high pitch – low pitch
tense – relaxed   smiling – sad
emotive – non-emotive*

Where do you think you should pause? Read the poem aloud.

2  Listen to the poem.

WRITING AND SPEAKING

1 Work in pairs. Make a list of romantic things, places and people.

Sunsets, Bilbao,...

2 Tell the rest of the class about the list you made in 1.

I once spent a romantic weekend in Bilbao.

3 Write a Valentine poem. You may like to use the words in the vocabulary boxes and the list you made in 1 to help you.

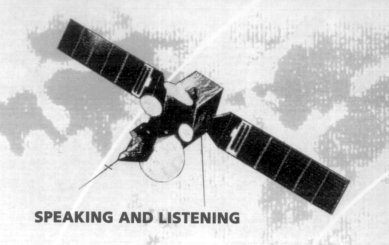

24 *Medium wave*

Reported speech (2): questions

VOCABULARY

1 This lesson is about the media – newspaper, television and radio. Look at the words in the box and put them into categories of your choice.

> advertisement article bias broadcast channel circulation commercial daily distribution documentary edition editor independent left wing liberal magazine national newspaper political popular programme publish quality reader readership regional right wing satellite show station video viewer weekly

2 Work in pairs. Do you have the same categories? Do you have the same words in the categories?

3 In the box are some words to describe different types of TV programmes or newspapers. Match them with their definitions below.

> broadsheet chat show quiz show soap opera tabloid sitcom

1 A television or radio comedy series in which the same characters appear in different stories each week.
2 A quality newspaper printed on a large sheet of paper.
3 A popular newspaper with lots of pictures, often in colour, and short articles; it's printed on a small sheet of paper.
4 A television or radio programme on which the presenter, or host, interviews well-known people.
5 A television or radio competition in which the players can win prizes if they answer questions correctly.
6 A television or radio programme about the continuing daily life and troubles of a group of characters, broadcast two or three times a week.

Which of your categories in 1 does each type of TV programme or newspaper go in?

SPEAKING AND LISTENING

1 Work in groups of three or four and talk about the media in your country. Answer the questions below.

1 Do people read mostly regional or national newspapers? ☐
2 Do most people read a newspaper every day? ☐
3 What are main newspapers? ☐
4 Which newspaper has the largest circulation? ☐
5 How many television channels are there? ☐
6 How much television do people watch every day? ☐
7 What are the most popular types of programmes? ☐
8 Is satellite television very popular? ☐
9 How many radio channels are there? ☐
10 Is radio more or less popular than television? ☐
11 What sort of programmes are there on the radio? ☐
12 Does the government control the media? ☐

2 🔲 Listen to Shelley, who's American, talking to Francis, who's English, about the media in the USA. Tick (✓) the questions above that Francis asks.

3 Work in pairs. Can you remember what Shelley's answers to the questions were?

🔲 Listen again and check.

GRAMMAR

> **Reported speech (2): questions**
> **You form reported questions by using *asked* +
> clause. You usually change the tense of the verb in
> the reported clause by moving it one tense further
> back into the past. The word order of the question
> always changes.**
>
> **Direct questions → Reported questions**
> **With question words (*who, what, how, why* etc)**
> *'Which newspaper has the largest circulation?' he asked.*
> *He **asked** which newspaper **had** the largest circulation.*
> *'How many channels are there?' he asked.*
> *He **asked** how many channels there **were**.*
>
> **Without question words**
> *'Is satellite television very popular?' he asked.*
> *He asked **if** satellite television **was** popular.*
> *'Does the government control the media?' he asked.*
> *He asked **if** the government **controlled** the media.*

1 Tick the statements which are true.

In reported questions, you:
1 use the auxiliary *do/does/did*.
2 change the word order.
3 use **if** for *yes/no* questions.
4 don't change the tense of the verb.

2 Turn these reported questions into direct questions.
Use *you*.

1 She asked who he was.
2 He asked if she read a newspaper.
3 She asked why he listened to the radio.
4 He asked if she watched a lot of television.
5 She asked if she had heard the news.
6 He asked what she was doing.

1 'Who are you?' she asked.

3 Turn these direct questions into reported questions.

1 'Do you read a daily newspaper?' she asked.
2 'What's your favourite TV programme?' she asked.
3 'Is satellite television growing?' she asked.
4 'Where do most people listen to the radio in your
  country?' she asked.
5 'Do you read a Sunday paper?' she asked.
6 'Have you watched any TV today?' she asked.

1 She asked if he read a daily newspaper.

WRITING AND SPEAKING

1 Look at the questions you ticked in *Speaking and
listening* activity 2. Write sentences saying which
questions Francis asked. Leave a gap between each
question.

*Francis asked Shelley if people read mostly regional or
national newspapers.*

2 Write a report of the interview with Shelley. Write a
brief summary of her replies under each question. Use
the linking phrases *the reason for this* and *because* to
explain the main reasons. Use the linking phrases
another reason for this and *besides* to explain further
reasons.

*She said that the regional press was more popular in
the USA. The reason for this was that the national
news was reported in the regional newspapers.*

3 Find out about other people's opinions of the media.
Go round the class asking questions about the
following:

– the newspapers which people read
– the amount of television they watch
– their favourite radio programmes

Think of supplementary questions about each
question.

Do you read a newspaper every day? Which one?

4 Tell the class what you asked people and what they
answered.

*I asked Giovanni if he read a newspaper every day.
He said he read 'La Stampa' most days.*

25 | *A cup of tea*

Reported speech (3): reporting verbs

SPEAKING AND READING

1 Work in pairs. The story in this lesson is *A cup of tea*, by a New Zealand writer, Katherine Mansfield (1888–1923), who lived in London for part of her life. Look at the picture and decide:

– where it takes place
– who the people are
– what's happening
– why it's called
 A cup of tea

2 Here are two parts of dialogue taken from part 1 of the story. Can you add anything to your answers to 1?

'Madam, would you let me have the price of a cup of tea?'
'A cup of tea? Then you have no money at all?' asked Rosemary.

'You're not taking me to the police station?' the girl stammered.
'The police station!' Rosemary laughed out. 'No, I only want to make you warm and to hear – anything you care to tell me.'

3 Read part 1 of the story and check your answers to 1 and 2.

Rosemary Fell was not exactly beautiful. Pretty? Well, if you took her to pieces... but why be so cruel as to take anyone to pieces. She was young, brilliant, extremely modern, exquisitely well-dressed. She was married and had a lovely boy. And her husband, Philip, absolutely adored her. They were rich, really rich.

One winter afternoon, she was buying something in a little antique shop in Curzon Street. The man showed her a little box, an exquisite enamel box. Rosemary liked it very much. 'Charming!' But what was the price? 'Twenty-eight guineas, madam.'

Twenty-eight guineas. Even if one is rich... 'Well, keep it for me – will you?'

The door shut with a click. Rain was falling, and with the rain it seemed the dark came too. Suddenly, at that moment, a young girl, thin, dark, shadowy – where had she come from? – was standing at Rosemary's elbow and said quietly,

'Madam, would you let me have the price of a cup of tea?'

'A cup of tea? Then you have no money at all?' asked Rosemary.

'None, madam,' came the answer.

'How extraordinary!' And suddenly it seemed to Rosemary such an adventure. Supposing she took the girl home? And she heard herself saying afterwards to the amazement of her friends, "I simply took her home with me," as she stepped forward and said, 'Come home to tea with me.'

The girl drew back startled. 'I mean it,' Rosemary said, smiling. And she felt how simple and kind her smile was. 'Come along.'

'You're not taking me to the police station?' the girl stammered.

'The police station!' Rosemary laughed out. 'No, I only want to make you warm and to hear – anything you care to tell me.'

Hungry people are easily led. The servant held open the door of the car and a moment later they were driving through the dusk.

'There!' said Rosemary. She had a feeling of triumph. She could have said, 'Now I've got you,' but of course she meant it kindly. She was going to prove to this girl that rich people had hearts, and that women were sisters.

GRAMMAR

> **Reported speech (3): reporting verbs**
> You can often use a reporting verb to describe the general sense of what someone thinks or says. This is common when you report orders, requests, warnings, threats, advice, invitations, offers and promises. There are several patterns for these verbs.
>
> | Pattern | Verbs |
> |---|---|
> | **1 verb + *to* + infinitive** | *agree, ask, decide, hope, promise, refuse* |
> | *Rosemary **decided to take** the girl home.* | |
> | **2 verb + object + *to* + infinitive** | *advise, ask, encourage, persuade, remind, warn* |
> | *Rosemary **persuaded the girl to come** home with her.* | |
> | **3 verb + (*that*) clause** | *agree, decide, explain, hope, promise, suggest, warn* |
> | *The girl **replied that** she didn't have any money.* | |
> | **4 verb + object + (*that*) clause** | *advise, tell, warn* |
> | *Rosemary **told the girl that** she wasn't taking her to the police station.* | |
> | **5 verb + object** | *accept, refuse* |
> | *She **accepted the invitation**.* | |
> | **6 verb + 2 objects** | *introduce, offer* |
> | *She **offered her a cup of tea**.* | |

1 Put the words in order to make reported statements.

1 her she come to along told
2 him she for keep told to it her
3 that she had none replied she
4 asked some money her to give she her
5 she tea invited home for her
6 her she promised she it meant

1 She told her to come along.

2 Underline the sentences in part 1 of the story which match the reported statements in activity 1.

1 'Come along.'

3 Here are some statements about part 2 of the story using reporting verbs. Decide who the pronouns refer to.

1 She invited her to sit down in a comfortable chair.
2 She told her not to be frightened.
3 She apologised and said she was going to faint.
4 She told her to come and get warm.
5 She told her not to cry.
6 She offered to look after her.

1 she – Rosemary, her – the girl

4 Write down what you think the people actually said.

'Please sit down in this comfortable chair.'

5 Here are some extracts from part 2 of the story. Rewrite them in reported speech using a suitable reporting verb.

1 'I'll arrange something, I promise.'
2 'Rosemary, may I come in?' It was Philip.
3 Philip asked, 'Explain. Who is she?'
4 'Look again. I think you're making a mistake.'
5 'Philip,' she whispered, 'am I pretty?'

1 She promised to arrange something.

6 Look at the reporting verbs in activities 3 and 5. Which pattern do they follow?

READING AND VOCABULARY

1 Work in pairs. Look at your answers to *Grammar* activities 3, 4 and 5. Can you guess what happens in part 2 of the story?

Now turn to Communication activity 9 on page 100 and read part 2 of the story.

2 Work in pairs. Try to remember as much as possible about the story. Use the words in the box to help you.

> adore antique shop bell
> boot box dazed desk
> dusk enamel faint
> pearls pretty prove
> scent servant smile
> softness startled tears
> warmth well-dressed

Now read the whole story again and check.

WRITING

Write a paragraph describing what happens in the story from the husband's point of view.

My wife was out shopping one day when she met a beautiful young woman in the street…

Progress check 21–25

VOCABULARY

1 In Progress check 16–20, you saw an explanation of the four types of multi-part, or phrasal verbs. Remember that many multi-part verbs are idiomatic and have several meanings. Here are some Type 1 multi-part verbs (which do not take an object).

break down	*The car broke down.*
take off	*The plane took off.*

Here are some Type 4 multi-part verbs (which have two particles).

run away with	*The thief ran away with my purse.*
face up to	*He faced up to his problems.*

Replace the words in italics in the sentences with phrasal verbs from the box.

> blow up   catch up with   do away with   drop off
> face up to   fall back on   fall through   get away
> get back   get down   get on with   give up
> go down with   go in for   go off   hold back
> look out   make up for   run away with   stand back
> stand up to   turn back   wear off

1 He *became ill with* pneumonia.
2 The enemy soldiers *surrendered*.
3 She couldn't do it alone, so she *relied on* him for support.
4 The milk *turned sour* in the sun.
5 I *have a good relationship with* her.
6 The bomb *exploded*.
7 The runner behind *drew level with* the one in front.
8 She *stopped* eating meat a year ago.

2 Look at the other multi-part verbs in the box in 1. Use a dictionary to check you know what they mean. Choose five verbs and write sentences showing their meaning.

3 Look at the vocabulary boxes in Lessons 21–25 again. Choose words which are useful to you and group them under headings of your choice in your *Wordbank*.

GRAMMAR

1 Choose the correct adjective or adverb.

1 It was an *incredible/incredibly* thrilling film.
2 He was *amazing/amazingly* rich.
3 The special effects were *spectacular/spectacularly*.
4 The action was rather *slow/slowly*.
5 The acting was really *funny/funnily*.
6 It was *terrible/terribly* depressing.

2 Complete the sentences with the adverbs and adverb phrases in brackets.

1 She moved ___ . (five years ago/to London)
2 He drove ___ . (down the street/slowly)
3 We were having lunch ___ . (one Sunday/in a restaurant)
4 She works ___ . (in the office/hard/all day)
5 He waited ___ . (for twenty minutes/at the bus stop)
6 Would you take this letter ___ ? (quickly/to the post office)

1 She moved to London five years ago.

3 Write these reported statements in direct speech.

1 He said it was raining.
2 She said she didn't have any money.
3 He said he liked watching television.
4 She said she would leave early.
5 He said he'd spoken to his mother last week.
6 She said she hadn't understood the question.

1 'It's raining,' he said.

4 Write these direct statements in reported speech.

1 'I'm late,' he said.
2 'I'm getting angry,' she said.
3 'She's gone,' he said.
4 'He's working abroad,' she said.
5 'I called her this morning,' he said.
6 'We'll pay for this,' she said.

1 He said he was late.

5 Punctuate this passage.

i was sitting in a restaurant when a man came in and sat down at the next table what can i get you sir the waiter asked him ill have what shes having he said and looked at me i was embarrassed and didn't know what to say he then asked me if i came here often i said this is the first time he then asked me why i was here on my own i said i was waiting for my husband to pay the bill then i got up and left

6 Rewrite the passage in 5 in reported speech.

7 Write reported questions.

1 'Who is he?' she asked.
2 'Why is she working so late?' he asked.
3 'Is she British?' he asked.
4 'What's her name?' he asked.
5 'What did she buy?' he asked.
6 'Will you have something to eat?' she asked.
7 'Have you got any money?' he asked.
8 'How did you do that?' she asked.

1 She asked who he was.

8 Rewrite the following sentences in reported speech using the reporting verbs below.

agree   decide   offer   promise   suggest   warn

1 'Would you like a lift?' he said.
2 'I know! I'll take the train,' she said.
3 'I'll ring when I get there,' he said.
4 'Don't touch that!' she said.
5 'Why don't you go and lie down?' she said.
6 'OK, I'll pay you ten pounds,' he said.

1 He offered her a lift.

SOUNDS

1 How do you pronounce the following words? Put them in four groups according to the way you pronounce the letter e: /iː/ /ɪ/ /e/ /ɜː/

welcome   cathedral   dense   pretty   complete   herbs   German

Listen and check.

2 Underline the silent consonant(s) in these words.

would   walk   answer   what   theatre   knock   night   neighbour   school   ought

Listen and say the words aloud.

3 Read this true sentence.

Peter asked where Paula came from and she said she was from London.

Listen and correct the statements below with the true sentence. Change the stressed word each time.

1 Peter asked where Maria came from and she said she was from London.
2 Peter asked where Paula came from and she said she was from Rome.
3 Peter asked where Paula lived and she said she was from London.
4 Philip asked where Paula came from and she said she was from London.

SPEAKING

1 Work in pairs. Find out what you have in common. Talk about:

– childhood   – education   – leisure interests
– favourite places   – holiday plans

2 Write two sentences about yourselves. Use different tenses.

Francesca and I are hoping to spend some time in Britain. We prefer summer to winter. We've never been to a foreign country.

3 Go round the class giving the information about yourselves. Make suitable responses to other students.

Francesca and I are hoping to spend some time in Britain.
Why don't you go there this summer?

Write down or remember the pieces of information that other students tell you.

4 Find your partner again and tell him or her what you learnt and what you replied.

Catherine said Francesca and she were hoping to spend some time in Britain. I suggested they go there this summer.

Eat your heart out... in the USA

Giving instructions and special advice

Anyone who thinks that food in the United States of America is all junk food will be astonished by the variety of the true cooking of the country. Scrapple, grand central oyster stew, jambalaya, tacos, cioppino and hashed browns are all American dishes, yet they come from different traditions and different regions of the country.

There are six main cooking regions in the USA: New England, New York, Deep South, Mid West, Tex Mex and West Coast. Here's a quick guide to what you can eat there.

★ In New England they eat a lot of fish and shellfish. Many dishes are left on the stove to be eaten all day, such as boiled beef and chicken stew, and Boston is the home of the famous baked beans.

★ New York is where people from all over the world meet, and you can see this in its cooking: Greek, Italian, Russian, Chinese and many others. Pizza and pasta are favourites, and it's the home of the hot dog and the hamburger.

Eat your heart out... in the USA

★ In the Deep South, it's a mix of English, French, African and Caribbean cooking, with spicy seafood, beans and rice, pork dishes, pecan pie and of course, southern fried chicken.

★ The farmland of the Mid West produces corn-on-the-cob (maize), steak, tomatoes, potatoes and lettuce, and baked hams. The people who live there came from Europe, so you can also try Hungarian goulash, Swiss, Dutch and English cheeses and Scandinavian coffee cake.

★ Tex Mex is hot and spicy, with green and red peppers, beans, tomatoes, mangoes, avocados, chocolate sauce (*mole*) and the fiery chilli con carne.

★ The West Coast is known for its fruit, especially oranges and lemons, and for its seafood, crabs, lobster and mussels. A lot of the cooking is with wine.

From brownies to tacos, from spare ribs to clam chowder, cooking in the USA has something for everyone. Enjoy!

VOCABULARY AND SPEAKING

1 Look at the words in the box and put them under these headings: *ways of preparing food, ways of cooking, food, kitchen equipment.*

> avocado bake beans boil bowl casserole cream
> chop cook cut dish fork fry frying pan
> green peppers grill ham hamburger heat hot dog
> knife mix mussels oil onion oven oysters pasta
> peel pepper pie pizza plate pour roast salt
> saucepan seafood slice spoon spread turkey

2 Work in pairs and talk about what dishes are typical of your country or region. What are typical ways of preparing food?

In Argentina, we eat a lot of meat, which we usually grill.

3 What food is typical to the United States of America?

READING AND LISTENING

1 Read *Eat your heart out... in the USA* and find food which is similar to food in your country. Which region would you like to visit? What would you like to try?

2 🔲 Listen to an American chef and find out where these dishes come from. There is one extra dish. Which one is it?

tacos   grand central oyster stew
hashed browns   jambalaya
scrapple   cioppino

3 Match the dishes the chef described in 2 with the ingredients.

1 pork, sausage, ham, prawns, tomato purée, onion, garlic, herbs and chilli, rice
2 tortillas, minced beef and pork, chilli and other spices
3 pork, onions, cornmeal, salt and pepper
4 fish, mussels, crab-meat, prawns, tomatoes, wine and herbs
5 oysters, butter, milk and cream

🔲 Listen again and check.

4 Here's the recipe for the extra dish, but the instructions are in the wrong order. Can you put them in the right order?

a Add the cream and cook the other side for about ten minutes.
b Check it is brown underneath then turn it over.
c Heat the butter and the oil in a frying pan. Mix the potatoes and the onions in a bowl.
d Add salt and pepper on the top and cook for about eight minutes.
e Spread the mixture over the bottom of the pan, and press flat.

🔲 Listen and check.

FUNCTIONS

> **Giving instructions**
> **You use the imperative or the present simple for giving instructions.**
> *Mix the potatoes and the onions and **cook** for ten minutes.*
> *You **mix** the potatoes and the onions and **you cook** for ten minutes.*
>
> **Giving special advice**
> ***Make sure you** use a heavy frying pan. **Always** spread the mixture evenly.*
> ***Don't forget to** add pepper and salt. **Never** heat the oil too much.*

1 Work in pairs. Ask and say what the dishes in *Reading and listening* activity 2 are and what they're made with.

What's scrapple?   It's a New England dish.

2 Write instructions how to make the following food and drink.

1 a cup of coffee   3 an omelette
2 a boiled egg   4 cheese on toast

3 Work in pairs. Here are some phrases which the chef used in his recipe. What special advice did he give? Use *make sure you, don't forget to, always,* and *never*.

use a heavy frying pan   add pepper and salt   press the mixture down
spread the mixture evenly   heat the oil too much   serve immediately

4 Think of special advice you can give for the food and drink in 2.

Never boil the coffee.

SPEAKING

1 Think about the typical cooking in your country and make a list of ten or twelve basic ingredients.

2 Show your list to another student. If he/she comes from the same country, do they agree with you? If he/she comes from another country, do you have the same ingredients?

3 Make a list of typical dishes from your region or country, and talk to your partner about them. Ask and say what they're made with, how you make them (if you know), and if there is any special advice.

Home thoughts from abroad

Defining relative clauses

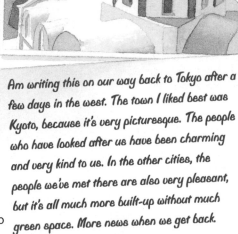

SPEAKING AND VOCABULARY

1 What sort of things do you write on your holiday postcards?

2 Which of these words do you associate with a good holiday?

> sunburn diving wreck mugging
> swimming pool delay mountain
> canal beach river island sunbathe
> sail postcard crowds ferry
> thunderstorm snow lost property
> speed limit lose your way danger
> visa customs dance peace

Can you use any of the words in the box to describe any holidays you've had? If the holiday has been disastrous, do you tell people about it?

READING

1 Read and match the two sides of the postcards.

2 Look at these words and phrases from the postcards. Who or what do the underlined words refer to?

1 <u>This</u> is the village where we're staying.
2 ... so <u>it's</u> beautifully decorated.
3 Will <u>tell</u> you all about <u>it</u> when we get back.
4 There are lots of amusing people <u>here</u>.
5 ... the people we've met <u>there</u> are also very pleasant.
6 ... but <u>it's</u> all much more built-up without much green space.

Am writing this on our way back to Tokyo after a few days in the west. The town I liked best was Kyoto, because it's very picturesque. The people who have looked after us have been charming and very kind to us. In the other cities, the people we've met there are also very pleasant, but it's all much more built-up without much green space. More news when we get back.

This is the village where we're staying. It's probably the most beautiful place we've ever seen, and very quiet. The people whose house we're staying in are interior designers, so it's beautifully decorated. There's a swimming pool which belongs to the house too. Will tell you all about it when we get back.

Having a lovely time. You can just see the hotel where we're staying, overlooking the sea. There are lots of amusing people here. The nicest person we've met is Spiros, whose brother owns the hotel. He's the man who does the cooking here. He's one of these people who are the life and soul of the party. We're having a great time.

GRAMMAR

> Defining relative clauses
>
> **You can define people, things and places with a relative clause beginning with *who, that, which, where* or *whose*.**
>
> **You use *who* or *that* to define people:**
>
> **– as a subject pronoun.**
>
> *He's the man **who/that** does the cooking at the hotel.*
>
> **In this sentence, *who* is the subject.**
>
> *He does the cooking.*
>
> **– as an object pronoun.**
>
> *The nicest person **who/that** we've met is Spiros.*
>
> **In this sentence, *who* is the object.**
>
> *We met **him**.*
>
> **You can leave out *who/that* when it is the object of the relative clause.**
>
> *The nicest person we've met is Spiros.*
>
> **You use *which* or *that* to define things:**
>
> **– as a subject pronoun.**
>
> *There's a swimming pool **which/that** belongs to the house.*
>
> **– as an object pronoun.**
>
> *It's probably the most beautiful place **which/that** we've ever seen.*
>
> **You can leave out *which/that* when it is the object of the relative clause.**
>
> *It's probably the most beautiful place we've ever seen.*
>
> **You use *where* to define places.**
>
> *This is the village **where** we're staying.*
>
> **You use *whose* to replace *his, her* and *their* in relative clauses.**
>
> *The people **whose** house we're staying in are interior designers.*
>
> **You can't leave out *where* and *whose* in relative clauses. Unlike non-defining relative clauses, you don't separate the relative clause from the main clause with a comma.**

1 Rewrite these sentences using a relative pronoun. There may be more than one possibility.

1 This is my friend. His house is for sale.
2 I spent two years in France. I learnt to speak French.
3 We went away with some friends. I work with them.
4 I lost the photos. I took them on holiday.
5 My sister works in Italy. Her company makes computers.
6 Why not spend a holiday in Recife? It's nice and warm there.

1 This is my friend whose house is for sale.

2 Which of the sentences in 1 can you rewrite leaving out the relative pronoun?

3 Look back at the postcards and find sentences where you can insert *who* or *which* as an object pronoun.

SOUNDS

1 A comma in written English often indicates a pause in spoken English. Put commas in the passage below.

This year I'm spending my summer vacation which is six weeks long in a small village near the Lakes which is very relaxing. The place where I'm staying only has about a hundred people who are mostly farmers although it gets crowded in summer when the people from Milan who have second homes in the region come and spend their holidays here too.

Now say the passage aloud. Make sure you pause at the commas.

2 🔲 Listen and check.

LISTENING AND WRITING

1 Read these questions and think about your answers to them.

Who is the most interesting or amusing person you've ever met on holiday?
Where is the most beautiful place you've ever been on holiday?
What is the most frightening thing that has ever happened to you on holiday?
What is the most exciting thing you've ever done on holiday?
Where is the most comfortable place you've stayed?

2 🔲 Listen and decide which questions in 1 these three speakers are answering.

3 Work in pairs. Try to remember as much as possible about what each person says.

🔲 Now listen again and check.

4 Imagine you have just been on holiday. Choose one of the questions in 1 and write a postcard describing the person, place or experience you've just had. Use the postcards in *Reading* to help you.

28 | *Local produce*

The passive

Scotland: It is thought that the most important single influence on the taste of Scotch is probably the Scottish water. This is why distilleries are often situated in narrow valleys near a stream. It is one of the few industries in Britain which will be encouraged to grow in the future.

Glasgow: Britain's third largest city, Glasgow, is famous for its shipbuilding. It was known above all as a port, but in the nineteenth century, the river was widened so that ships could be built on its banks. But the industry has been affected by strong foreign competition and the number of ships under construction is being reduced.

Staffordshire: Pottery is made in this region, and the beautifully decorated Wedgewood, Minton and Spode china is collected by people all over the world. When the canals were built in the eighteenth century, the raw materials, such as clay, were brought to the region from the west of the country and china was exported all over the world.

Yorkshire: Sheep have been bred on the Yorkshire moors for their wool for centuries, and the streams have provided water for the mills, and for washing and preparing the wool.

South Wales: Until a few years ago, coal was still mined here, but during the 1970s and 80s the industry was being increasingly affected by the use of alternative sources of power, such as nuclear electricity, and mining was stopped completely in the 1990s.

Hereford: This is the home of the world's largest cider factory. Apples are brought from the surrounding orchards, processed by the factory and distributed all over the country.

Sheffield: Steel and cutlery were first manufactured in Sheffield because it was near to the raw materials and resources which are needed for the industry: forests, streams and iron ore.

VOCABULARY AND READING

1 Underline the verbs in the box below.

bananas beer build cars china cloth coal coffee computers cotton design electricity factory field fruit gold grow invent iron make manufacture mine pottery power station produce rice ships steel tea tobacco wheat wine wool workshop yard

Think of nouns which come from them.

build – building

2 Match the verbs you underlined in 1 with the nouns in the box. (Some can go with more than one noun).

build – factory,...

Can you describe each phrase as a *natural* or a *manufacturing* process?

3 *Local produce* is about some of the things which are made or grown in Britain. Read it and find out if Britain has more natural produce or manufactured produce.

4 Look at these words from the passage which may be unfamiliar to you. Decide if they are nouns or verbs. If they are nouns, decide if they are likely to be a *manufactured product*, a *natural product* or a *place*.

clay   cutlery   iron ore   bred   mill   stream   distillery   cider   orchard

5 Does your country produce the same things as Britain? If so, write down where they are produced. If not, from which countries are they imported?

GRAMMAR

> The passive
>
> **You form passive verbs with the different tenses of *be* + past participle.**
>
> **You use the passive:**
>
> **– when you want to focus on when or where something is done, or what is done, rather than who does it.**
>
> *The valleys of Wales **are linked** with the coal industry.*
>
> **– to introduce general opinions.**
>
> *It **is thought** that the most important single influence...*
>
> **– to describe processes.**
>
> *Apples **are brought** from the orchards, **processed** and **distributed**.*

1 Look at the passage again and underline all the passive verbs. What tense are they in?

2 Complete these sentences using a suitable passive tense of the verb in brackets.

1 Gunpowder ____ in China. (discover)
2 Scotch whisky ____ all over the world. (export)
3 Today, shipbuilding ____ by competition from the Far East. (affect)
4 Wine ____ in France for centuries. (make)
5 In the next century, more whisky ____ from Scotland than wine from France. (export)

1 Gunpowder was discovered in China.

3 Rewrite these sentences in the passive.

1 China makes three times as many bicycles as the USA and Japan.
2 Exxon, a giant oil company in New York, earns more money in a year than many countries.
3 The Romans first mined coal in the first century AD.
4 The Dutch grow about 3,000 million flowers a year.
5 An average American uses twice as much fuel as an average European.

4 Look at these pairs of active and passive sentences. In each pair, which do *you* think is the better sentence? Explain why.

1 a We only use 11 per cent of the Earth's land for farming. But each year we use less land because rain washes the soil away.
   b Only 11 per cent of the Earth's land is used for farming. But each year less land is used because the soil is washed away by the rain or blown away by the wind.

2 a The Japanese like fish and eat 3,400 million kilograms of it a year.
   b Fish is liked by the Japanese and 3,400 million kilograms of it is eaten by them each year.

3 a At first, the Europeans didn't smoke tobacco but used it as an ornamental flower.
   b At first, tobacco wasn't smoked by the Europeans but used as an ornamental flower.

SPEAKING AND WRITING

1 Work in pairs. Is there a particular food or drink or a product which your town, region or country is famous for? Make notes about it. Think about:

– how long it has been made or grown there
– when it was first made or grown
– how it's made or grown
– how it is processed
– who it is bought or used by
– what changes will be made in the future

2 Go round the class and find out about other people's local produce, and take notes.

3 Use the notes you took in 2 and the passage in *Vocabulary and reading* to write a description of some of the local products of people in your class.

Coffee has been grown in Brazil for hundreds of years.

Just what we're looking for!

Verb patterns (2): *need + -ing* and passive infinitive; causative construction with *have* and *get*, reflexive pronouns

29 June

I left the hotel today at eight o'clock for an early appointment with the agent, who yesterday assured me he has found me just the house we're looking for. When I arrived he looked less convinced than I did. Even at that time of the morning, I was already irritable and despondent as I arrived at the first address. At first I walked past the house. Where it should have been was a wilderness of trees and overgrown grass. Then out of the green darkness stepped the agent. 'Ah, there you are! It's here,' he said. I stepped in through the broken-down gate, and walked up the dusty garden path. It immediately felt cooler and calmer. The agent rattled a large bunch of keys, and tried several in the door, talking to himself all the time, before he exclaimed, 'Ah!'

We let ourselves in and walked into a deliciously cool, but dusty house. He suggested I walk around by myself. I went into a gloomy living room downstairs and switched on the light, but nothing happened. The agent heard me clicking the switch and said, 'Ah! No good. The switch needs mending. I'll have an electrician repair it immediately.' I peered into the darkness and made out the shape of a window on the far side. The agent walked over to the window and threw open the peeling shutters, and the sunlight streamed in. A rather faded sofa and two battered armchairs sat around an open fireplace which hinted at log fires in winter. The curtains were stripy but more or less in shreds. Outside there was a terrace and beyond the trees, the dense undergrowth, the tall grass and the wild plants that were once the garden, were the mountains in the distance.

I walked through into what must have been a kitchen, but only recognisable because of the antiquated equipment which I last saw during a visit to a local museum. I turned on the tap, and once again, nothing happened. 'The water needs to be reconnected. We'll get the plumber to do it. It's no problem,' the agent said.

Upstairs there were two bedrooms and a bathroom with low ceilings and which were, despite being hidden in the roof, still quite cool. The bathroom had no bath and not much room, but a beautiful view over the garden. The basin was filthy with the dirt of the years during which the place had been unoccupied. I sat on the brass bed in the dusty bedroom, and looked round, thinking, not bad, not bad at all. In my mind, I could see the house with new curtains and carpets, our own furniture, which had been in store for several months, books on the shelves, beds made up, lengthy lunches on the terrace, endless summers and warm winters. I could do most of the work myself.

I went downstairs and the agent looked at me hopefully. It was worth the lengthy search, the dusty visits to endless houses, the depressing inspections of grim flats, to see his smile when I said to him, 'It's just what we're looking for.'

VOCABULARY AND READING

1 You can use the adjectives in the first box to describe the features of a house which have something wrong with them.

> blocked torn worn faded dirty broken dusty
> not working overgrown overflowing dripping stuck
> scratched stained old-fashioned burst peeling

What features of a house can you use them to describe? Match the adjectives with the nouns in the box below.

> tap shower oven carpet toilet armchair sofa
> table curtains garden light dishwasher gate stairs
> sink bath lawnmower shutters wallpaper ceiling
> wall window pane wiring

blocked: sink, bath …

2 Which of these words go with the words in 1?

> builder carpenter clean cut decorator electrician
> gardener fix mend paint plumber repair replaster
> replace wash

3 The passage comes from the diary of a man looking for somewhere to live. Read it and decide if the event he describes comes from the beginning, middle or end of his search.

4 Look at these words from the passage which may be new to you.

> battered in shreds dense antiquated filthy

Can you guess what they mean? Choose a word from the first box in *Vocabulary and reading* which has a similar meaning.

5 Read the passage again and make notes about what is wrong with some of the features of the house.

gate – broken

GRAMMAR

> *Need + -ing* and passive infinitive
> You can use *need* + an *-ing* verb form or a passive infinitive to say what it is necessary to do.
> *The switch **needs mending**.*
> *The switch **needs to be mended**.*
> *The water **needed reconnecting**.*
> *The water **needed to be reconnected**.*
>
> Causative constructions with *have* and *get*
> You can use *will have* + infinitive and *will get* + infinitive to mean *ask* or *order* someone to do something.
> *I'll **have** an electrician **repair** it.*
> *We'll **get** the plumber to **do** it.*
>
> Reflexive pronouns
> *me – myself, you – yourself/yourselves, him – himself, her – herself, it – itself, we – ourselves, they – themselves*
> *I'd do most of the work myself.*

1 Read the passage again and say what needs doing or needs to be done.

The gate needs mending.

2 Say who you ask if you need to have someone do, or to get someone to do the following things.

1 mend a gate   4 reconnect the electricity
2 cut the grass   5 paint the shutters
3 clean the house   6 reconnect the water

3 Write sentences saying which things in 1 or 2 you would do yourself, and what you'd get someone to do.

I'd get a carpenter to mend the gate.
I'd cut the grass myself.

SPEAKING

1 Work in pairs. Are you houseproud? Do you care if your home isn't in perfect condition? Say what's wrong with it at the moment.

The shower isn't working. The oven is dirty.

2 Find out what needs doing in your partner's home. Is your partner going to do it him/herself? Or will he/she get someone else to do it?

3 Work in groups of four or five. Prepare a description of a home that would suit all of you.

30 | *Sporting chance*

**Verb patterns (3): *make* and *let*;
infinitive constructions after adjectives**

VOCABULARY

1 Look at the words in the box. Which
words are sports?

athlete basket compete basketball
boxing swimming goal play
referee skiing serve pool ring
football player fight court field
slope boots racket match game
tennis run athletics boxer score
race lane send off

basketball,...

2 Which of the other words in the box
go with the sports? (Some words can
go with more than one sport).

READING AND LISTENING

1 Work in pairs. What's your favourite
sport? Why do you enjoy it? Is there
anything you dislike about it?

2 Read *Sporting chance* and decide
which sports the suggestions
refer to.

3 🔲 Listen and check your
answers to 2.

> **Watching and playing sport**
> is still one of the world's most popular
> leisure activities. But with the arrival of
> professional competition, many people think some
> sports, like tennis and football, have changed and have
> become too predictable. What's your favourite sport? Do you think
> it has become boring, too dangerous or even too safe? Here are some
> suggestions about how some sports could be improved to give
> players and spectators a sporting chance.
>
> **1** 'I would make them use a circular pool and make them try to pass each other.'
> **2** 'I would let the referee look at a video recording before he
> decides to send a player off.'
> **3** 'I wouldn't let people over two metres tall play.'
> **4** 'Why don't they make them fight in a round ring, not in the traditional square?'
> **5** 'I wouldn't let the players have a second serve.'
> **6** 'I'd let them all go down the slope at the same time,
> racing against each other.'
> **7** 'If the goal posts were wider apart it would let players score more.'
> **8** 'Make the players use heavier or wooden rackets.'
>
> Adapted from *Breaking the rules*
> by Norman Harris, *The Observer*.

4 Here are the reasons the speakers give. Match them with their suggestions.

a 'It's very easy for the players to score, so it isn't very exciting to watch. If the players were shorter or the basket higher, it would make them work harder.'

b 'At the moment they're quite narrow, and it's difficult for players to score very often in a match.'

c 'It's essential to slow the game down. We want to see how well they can play, not how strong they are.'

d 'It would be easier to see that they are competing against each other.'

e 'It's boring to watch them, one after the other, racing against the clock and not against each other.'

f 'It's likely to need more accuracy and less power.'

g 'It's quite usual for a fighter to get trapped in the corner. If the ring was round, it would be safer.'

h 'It's very hard to make up your mind if you haven't seen exactly what happened.'

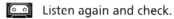 Listen again and check.

GRAMMAR

> ### Verb patterns: *make* and *let*
> **You can use *make* + noun/pronoun + infinitive to express an obligation.**
> *Make the players use* heavier rackets.
> *Why don't they make them fight in a round ring?*
> **You can use *let* + noun + infinitive to express permission.**
> *I would let the referee look at a video recording.*
> **You can use *not let* to express prohibition.**
> *I wouldn't let the players have a second serve.*
>
> ### Infinitive constructions after adjectives
> **You can use an infinitive (with *to*) after these adjectives:**
> *boring, cheap, dangerous, difficult, easy, essential, expensive, good, hard, important, interesting, (un)likely, (un)necessary, (im)possible, (un)pleasant, right, safe, (un)usual, wrong*
> *It's essential to slow the game down.*
> **You can use *for* + noun/pronoun between the adjective and the infinitive.**
> *It's very easy for the players to score.*

1 Complete these sentences with the correct form of *make, let* or *not let*.

1 Would you ___ me take you to the match?
2 They ___ you touch the ball in football.
3 They ___ you have two serves in tennis.
4 They ___ him swim in a freezing pool.
5 They ___ you stay in your seats in stadiums in Britain.
6 Do they ___ you touch the ball with your head?

2 Think of a sport you play or enjoy watching. What would you do to improve it? Use *make, let* and *not let*.

In tennis, I would make the players use wooden rackets. I wouldn't let them use metal frame rackets.

3 Rewrite these sentences with an infinitive construction.

1 Watching football is very exciting.
2 Slowing the game down is essential.
3 Getting tickets is sometimes difficult.
4 For girls, playing football is unusual.
5 Scoring goals is getting very easy.
6 Being interested in sport is good for children.
7 Tennis is a pleasant game to play.
8 Doing sport professionally is expensive.

1 Football is very exciting to watch.

SOUNDS

Underline the stressed syllables in these sentences.

I would make them use a circular pool and make them try to pass each other, so that it's more like a running race. It would be easier to see that they're competing against each other.

I would let the referee look at a video recording before he decides to send a player off. It's very hard to make up your mind if you haven't seen exactly what happened.

 Now listen and check.

SPEAKING

1 Read these statements and decide if you agree with them.

It's essential to make young people play sport.
It's good to encourage children to play a competitive sport.
It's wrong to let players in any sport receive money for playing.
It's right to ban sports men and women who use drugs.
It's unnecessary to change the rules of a sport.

2 Find out what other people in your class think about the statements.

Progress check 26–30

VOCABULARY

1 Some of the most common structures in everyday English are verb + noun combinations. Most of the meaning is contained in the noun. If you use them you will sound more fluent.

have have a bath   **take** take a photo
do   do the shopping   **make** make a speech
give give someone a ring

How many structures can you make using the verbs above and the following nouns? Some verbs can go with more than one noun.

an opinion   your hair   time   help   a mistake
your best   prisoners   visitors   a lecture   a cold
the bed   a party   a taxi   room   a move   a note
the ironing   a point   a test   advice   a smile
an example   the washing   damage

2 Complete these sentences with a suitable form of *have, make, take, give* or *do*.

1 Can you ___ space for me on this seat?
2 Have you ___ the washing-up yet?
3 He's ___ a really bad cold since last week.
4 If you ___ the train, it will be cheaper but slower.
5 We're ___ a dinner party next Sunday.

3 There are many words in English which are sometimes confused. Here are some of them.

come – go, buy – sell, borrow – lend, lie – lay,
take – bring, sensible – sensitive, check – control,
now – actually, expect – wait, watch – see,
listen – hear, loose – lose, economic – economical,
look – watch, forget – leave

4 Choose the correct word in these sentences.

1 The government revealed its new *economic/economical* strategy.
2 He was very *sensible/sensitive* to criticism.
3 Can I *borrow/lend* you some money?
4 Can I *check/control* your ticket to see if you've got the right one?
5 I *lay/lied* down on the bed.

5 Complete the sentences with pairs of words from the list in 3.

1 I was ___ to a CD when I ___ a noise outside.
2 When you ___ weight, your trousers are ___ .
3 When you come to our party, ___ a taxi and ___ a bottle.
4 I'm ___ for a bus. I ___ one any minute now.
5 ___ , I want to come and see you right ___ .

6 Look at the vocabulary boxes for lessons 26–30 again. Choose words which are useful to you and write them in your *Wordbank*.

GRAMMAR

1 Write detailed instructions on how to do one of the following things. Use infinitives, imperatives and *always, never, make sure you, don't forget to.*

1 earn lots of money
2 make a cup of coffee
3 stay healthy
4 spend money quickly
5 learn English

1 To earn lots of money, make sure you always work hard, and never take any holiday.

2 Rewrite these sentences using a suitable relative pronoun.

1 This is the car. It's for sale.
2 She went to the USA. She met her husband there.
3 On holiday he met a friend. He knew her from home.
4 Someone stole the bag. She bought it in Paris.
5 She lives with a friend. His family owns a large house in the country.

1 This is the car which is for sale.

3 Which sentences in 2 can be rewritten leaving out the relative pronoun?

4 Complete the sentences using a suitable passive form of the verb in brackets.

1 Mercedes cars ___ in Germany. (make)
2 Skiing ___ in Norway. (invent)
3 Champagne ___ in France for centuries. (produce)
4 In the future more ships ___ in the Far East than in Europe. (build)
5 It ___ that smoking is bad for your health. (recognise)

1 Mercedes cars are made in Germany.

5 Write sentences saying what needs doing. Use the words in brackets.

1 My car has broken down. (take to garage)
2 We've got a burst pipe. (mend)
3 The wallpaper is peeling. (redecorate)
4 The grass is very long. (cut)
5 The electricity has been cut off. (reconnect)

1 It needs taking to the garage.

6 Say which jobs in 5 you'll do yourself and which jobs you'll get someone to do.

7 Say what Jim's employers *make* him, *let* him or *don't let* him do.

1 I have to arrive at eight-thirty.
2 I'm not allowed to go home for lunch.
3 I can go home at four in the afternoon.
4 I have to work on Saturdays.
5 I can have six weeks holiday.

1 They make him arrive at eight-thirty.

SOUNDS

1 How do you pronounce the following words? Put them in two groups according to the way you pronounce the letter i: /aɪ/ or /ɪ/.

bite bit dish knife pie pick olive iron rice light

🔊 Listen and check. Say the words aloud.

2 Say these words aloud. Is the underlined sound /ʃ/ or /dʒ/? Put the words in two columns.

pre<u>c</u>ious stran<u>g</u>er chan<u>g</u>e suspi<u>c</u>ious passen<u>g</u>er fini<u>sh</u> <u>sh</u>ower <u>sh</u>are su<u>gg</u>est dan<u>g</u>er <u>j</u>ob

🔊 Listen and check.

3 Match the words with their phonemic transcription.

1 /haɪ/	2 /daɪ/	3 /beə(r)/	4 /deə(r)/	5 /waɪ/	6 /weɪ/
7 /deɪ/	8 /weə(r)/	9 /beɪ/	10 /heə(r)/	11 /heɪ/	12 /baɪ/

why	wear	way	buy	bear	bay
high	hair	hay	die	dare	day

🔊 Listen and check. Say the words aloud.

READING AND WRITING

1 Read the information about inventions and decide where these sentences go.

a It was powered by steam and looked like a giant kettle.
b It is reported that a stream of customers began to use the machine again and again, feeling faint each time.
c Speed checks were set up in 1902.

1 The first car was invented by Nicholas Cugnot in France. On its trial-run it ran perfectly for the first few minutes. But when Cugnot increased the speed, he lost control and crashed into a wall.

2 People and horses in England were frightened by the motor car when it was first introduced. It was considered to be noisy, dangerous and dirty. Policemen hid behind hedges to catch drivers who were going too fast.

3 When the first escalator was installed in Harrods, the department store in London, the ride was considered to be dangerous by some people. So the management decided to serve brandy free of charge to any passengers who felt faint when they reached the top. It soon became a success.

2 Here are some words which can go in the passages:
motor, stone, soon. They go:
The first **motor** car crashed into a **stone** wall. He **soon** lost control

Think of some more words which can go in each passage. Write them on a separate piece of paper in the wrong order.

3 Work in pairs and exchange words. Decide where and in which passage your partner's words can go.

31 *I never leave home without it*

Zero conditional; *in case*

VOCABULARY AND LISTENING

1 Which of the following items do you have in your bag or pocket at the moment?

> alarm clock credit card business card cheque book cash ballpoint pen binoculars mirror address book notebook paperback novel penknife map chocolate keys mobile phone diary umbrella handkerchief

2 🔊 Listen to eight people talking about what they have in their bag or pocket at the moment. Put the number of the speaker by the items in 1. (Some of them mention more than one item.)

3 Work in pairs and check your answers to 2. Can you remember why each speaker never leaves home without the items they talk about?

4 🔊 Listen again and check your answers to 3.

GRAMMAR

Zero conditional

You can follow *if* with a number of tenses. When you talk about something which is generally true or which usually happens, you can put a present simple or a modal after the *if* clause. This is sometimes called the zero conditional.

*If there's a long wait, I **take** something to read.*
*If I take a spare key, I **don't have to** disturb the neighbours.*

In case

You use *in case* to give the reason why you do something to be ready or safe for something that might happen. You use the present simple after *in case* but not a future tense or *going to*.

*I always carry a bar of chocolate **in case** I get hungry.* (**I carry some chocolate with me all the time because there is a chance that I'll get hungry.**)

You can also put a past tense after *in case.*
*She took a sweater **in case** it **got** cold.*

You can use *if* instead of *in case* when it's more certain that something else will happen.

*I always take a book **if** I have to wait for a bus.*
(**If I know I will have to wait, I take a paperback.**)
*I buy a bar of chocolate **if** I get hungry.*
(**When I get hungry, I always buy some chocolate.**)

1 Complete these sentences with *in case* or *if.*

1 Can you collect me from the station ____ the train arrives late?
2 I'll take something to drink ____ I get thirsty.
3 I'll arrive with plenty of local currency ____ the airport bank is closed.
4 Can you get me some stamps ____ the post office is open.
5 He got to the station early ____ he missed the train.
6 Use my pen ____ yours doesn't work.

2 Write your answers to *Vocabulary and listening* activity 3 using *in case.*

He always takes a spare key in case he leaves his keys at home.

3 Choose six more items from the box and write sentences explaining why it's a good idea to take them.

It's a good idea to take an umbrella in case it rains.

SOUNDS

Underline the stressed syllables in these words.

alarm business binoculars address penknife chocolate umbrella paperback novel handkerchief

🔊 Listen and check. Say the words aloud.

READING

1 You're going to read a passage by John Hatt, an experienced traveller, about the things he never leaves home without. Can you decide why he might take the following things?

torch   maps   neck pillow   binoculars
insect repellent   door wedge   earplugs
teaspoon   light bulb

2 Read the passage and match the items in 1 with the descriptions.

3 Answer the questions about difficult vocabulary.

trapped - Is this likely to mean that in the dark without a torch you can or can't escape? (paragraph 1)

banned - If books are difficult to find are they likely to be legal or illegal? (paragraph 2)

minimal - Is it likely to take up a lot or very little room? (paragraph 4)

unscrewed - Does this mean something like take out or put in the old bulb? (paragraph 6)

4 Work in pairs. Can you remember why he takes each item? Read the passage again and check.

He takes earplugs in case there is a lot of noise.

SPEAKING

1 Work in pairs. Talk about equipment you usually take on a car journey, a skiing trip, or a walking holiday. Explain why you take each item using the zero conditional, *in case* and *if*.

2 Is there anything you never leave home without? Make a list of the things you have in your handbag, briefcase or wallet. Which items are essential?

1   In certain places it becomes an everyday tool, and in case of emergencies it is essential. After the bomb explosion at the Grand Hotel in Brighton, Mrs Thatcher started a habit of keeping one by her bed. She had discovered that you are trapped in the dark without one.

2   These often can't be bought after your journey has begun. Even where there are bookshops, buy them before you go in case the best ones are out of stock or politically unacceptable and banned.

3   Although they are an entirely unnecessary piece of equipment, I always travel with them. I then know that I stand a chance of getting some sleep. In much of the world, you may be obliged to sleep against the background of a television, juke box and tape recorder all at full volume.

4   It takes up minimal room and is useful in case you want to eat snacks during journeys or in a hotel room. If you're equipped with a penknife as well, you can eat almost anything.

5   They are useful in cheap hotels, in case you can't lock the door from the inside; sleeping friends have been robbed by thieves entering through the bedroom door.

6   On my first evening in Cuba I had dinner with a friend who had just spent three weeks there. She gave me one, claiming that it was one of the best presents I would ever get. She was correct. Every evening I unscrewed the miserable, dim one in my hotel room and replaced it with the 150 watt one. This cheered up the room and, more important, enabled me to read.

7   If you don't get the window seat on a train, bus or plane, falling asleep can be uncomfortable: when nodding off, your head suddenly lolls uncomfortably to one side. In recent years blow-up ones have been widely marketed, and can be found in big stores or airport shops. With one of these you can sleep comfortably on any transport.

Describing a sequence of events (3):
as soon as, *when* and *after* for future events

A

Mr H Fish
203, Station Approach
Henley on Thames
Oxfordshire

Clarke's Garage
Frenchay Road
Reading
Berkshire

Dear Mr Fish,
Thank you for your letter of ...
We are sorry to hear that your recent purchase has already developed faults. We will send a mechanic to collect the car when we are less busy and have received the parts. This is likely to be in a week or two.
Yours sincerely,

Patrick Cl

Patrick Clarke

B

Clarke's Garage
Frenchay Road
Reading
Berkshire

Ward B, Henley Hospital
Henley on Thames
Oxfordshire

Dear Mr Clarke,
Further to our conversation about my car, I am writing to inform you that there is now no need to collect it.
While I was driving to the local garage to have it mended at my own expense, the brakes failed, the steering wheel came off in my hands, and the engine stopped. I drove into a brick wall, and am recovering in hospital. I am expected to leave hospital in three weeks' time.
I would be grateful if you would refund the money I paid for the car as soon as you receive this letter.
Yours sincerely,
Hector Fish

C

Clarke's Garage
Frenchay Road
Reading
Berkshire

203, Station Approach
Henley on Thames
Oxfordshire

Dear Sir or Madam,
I am writing to you about the car I bought from you on 7th July.
On the journey home, the speedometer stopped and the headlights didn't work. I am aware that it was a second-hand car, but feel that these essential features should be in working order.
As I am unable to return the car to your garage, I would be grateful if you would make arrangements to collect it as soon as you receive this letter.
I look forward to hearing from you.
Yours faithfully,
Hector Fish

D

Mr H Fish
Ward B, Henley Hospital
Henley on Thames
Oxfordshire

Clarke's Garage
Frenchay Road
Reading
Berkshire

Dear Mr Fish,
I am writing to send you my best wishes for a speedy recovery from your recent injury.
As you now no longer have a car, I wonder if you would be interested in another of our second-hand models. We have a superb selection of cars in perfect condition, and will be pleased to discuss the matter with you after you have recovered.
I look forward to seeing you then.
Yours sincerely,

Patric

Patrick Cl

E

Clarke's Garage
Frenchay Road
Reading
Berkshire

203, Station Approach
Henley on Thames
Oxfordshire

Dear Mr Clarke,
I am writing again about the car which I bought from you on 7th July.
You have still not sent anyone to collect the car. In the meantime, apart from the faulty speedometer and the headlights, the police stopped me recently because the indicators aren't working, and the seat belts have jammed. I am now expecting a fine.
I would be grateful if you would tell me why you haven't sent anyone to collect the car. I must insist that you do so immediately.
Yours sincerely,
Hector Fish

SPEAKING

1 Work in pairs. How often do you have to make a complaint about something? Think of situations when you have made a complaint.

2 Which of the situations you thought of in 1 do you complain about in writing? Which do you think is more effective: a letter or a face to face meeting? Do you think it is better to make your complaint angrily or politely?

VOCABULARY

1 Work in pairs. Which of these parts of the car can you see in the picture?

> accelerator aerial bonnet boot
> brake brake light bumper clutch
> cassette player dashboard engine
> exhaust fuel gauge gear lever
> headlight indicator jack mirror
> number plate pedal radio
> rear light roof seat belt sidelight
> speedometer steering wheel
> wheel windscreen wing wipers

2 Write simple definitions for the parts of the car in the box which you can't see in the photo.

The brake light is the light that comes on when you brake.

READING AND LISTENING

1 The letters between Hector Fish, who has just bought a car, and Patrick Clarke, who sold it to him, have been printed without dates. Read them and match the dates with the letters.

> – 10 July   – 14 July   – 16 July
> – 25 July   – 30 July

2 Listen to a telephone conversation between the two men and decide after which letter the conversation took place.

GRAMMAR

> **Talking about the future**
> **You can use the following expressions to talk about the future.**
> **– for something that is arranged.**
> ***to be due to* + infinitive**
> *We **are due to receive** the parts in a week's time.*
> **– for something that may happen but is not certain.**
> ***to be likely to, to be expected to* + infinitive**
> *This **is likely to be** in a few days. I **am expected to leave** hospital in three weeks' time.*
>
> **Describing a sequence of events in the future**
> **You use the present simple or present perfect to express the future in time clauses after *as soon as, when* and *after*.**
> *Please make arrangements to collect it **as soon as** you **receive** this letter.*
> *We will send someone **when** we **are** less busy.*
> *We will be pleased to discuss it with you **after** you **have recovered**.*
> ***As soon as* means immediately. *When* and *after* are less definite.**

1 Write sentences describing three things which are due to happen to you in the next two weeks and three things which are likely/expected to happen.

2 Write sentences beginning with *when, as soon as* and *after* in turn.

1 the spare parts arrive/we ring you
2 we are less busy/we contact you
3 you receive this letter/you collect the car
4 you leave hospital/you come to our showroom
5 you come to our showroom/we show you a new car
6 I get better/I go to see my lawyer

1 When the spare parts arrive, we'll ring you.

WRITING

1 In what order do you expect to see these features in a letter of complaint?

name and address of person you're sending it to   your signature and name request for information or action   date   reason for writing   conclusion your address   complaint   closing remarks   greeting

Look at the letters and see if you can find the features in each one.

2 How do you start and finish a letter to someone whose name:
a you know   b you don't know?

3 Write a letter of complaint about one of the following:

– a new camera which doesn't work   – a faulty television
– a book which has fallen apart   – bad service in a restaurant

Use the features in activity 1 and the letters to help you.

33 *Superhints*

**Verb patterns (4): infinitive of purpose;
by + *-ing*; giving advice: *if* clauses**

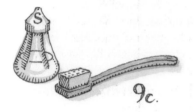

fig.7.

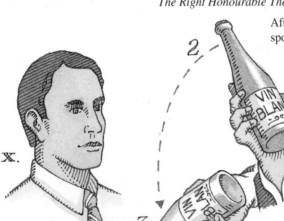

Superhints **is a book of good ideas: a collection of helpful suggestions on a variety of subjects – from cookery, gardening and DIY to travelling, planning ahead and emergencies – contributed by judges, TV personalities, actors, designers, politicians, beauty experts, company executives and housewives.**

To make your windows really shine, clean them with wet newspaper, then polish them with a soft cloth.
Dame Judi Dench, Actress

To keep teeth shiny and bright, brush them occasionally with salt.
Mrs Nigel Talbot-Rice

To clean flies off the windscreen, put toothpaste on with a wet rag, wash it off and polish the glass with newspaper.
Rosemary, Marchioness of Northampton

If a lettuce becomes limp, place it in an empty saucepan with a lump of coal. Put the lid on, and after a few hours it will become crisp again.
Mrs Darling's nanny

When, at a meal, soup or food gets on a tie, soak it well with dry white wine – this usually takes the stain out.
The Right Honourable The Lord Havers

After removing a stain it helps to avoid a ring if you dry the damp spot quickly with a hairdryer.
Dorothy Parker, Writer

If an oven dish gets burnt, put it face down on the grass all night: next morning it will easily wash clean.
Field Marshall, The Lord Bramall

To comfort a young puppy, put a ticking clock in its basket, along with a hot-water bottle, both wrapped in a blanket.
Mrs Lionel Green

To test if an egg is fresh, place it in a bowl of water. If the egg floats, it is bad.
Nigel Talbot-Rice

Fresh coffee, once opened, soon loses its aroma. By adding a cube of plain chocolate to coffee just before using, you can restore the aroma. A teaspoon of cocoa powder would do, but chocolate is best.
Roland Klein

If red wine gets dropped on the carpet, pour white wine on it immediately, and leave it for five or ten minutes before mopping it up.
HRH The Princess Margaret, Countess of Snowden

Adapted from *Superhints*, compiled by The Lady Wardington

VOCABULARY AND READING

1 Look at the words in the box. Underline the verbs.

aroma ash basket blanket boiling water bright brush off carpet chewing gum clock coal crisp dish fly hiccups ice cube lettuce lid limp lump mark mop up olive oil oven polish puppy rag remove scrape off shine shiny sip spot spread stain stretch tie tight wet wrap

78

2 Work in pairs. Do you know how to do these things?

- make lettuce crisp again
- avoid rings after removing a stain
- make your windows really shine
- remove red wine from the carpet
- clean a burnt oven dish easily

3 Read the *Superhints* and find the answers.

4 Work in pairs. Which is the most useful hint for you? Which is the least useful?

GRAMMAR

> **Infinitive of purpose**
> You use *to* + infinitive to say how you do something.
> *To make* your windows really shine, **clean** them with wet newspaper.
> You can also put the clause of purpose at the end of the sentence.
> *Clean your windows with wet newspaper* **to make** them really shine.
>
> **By + -ing**
> You can use *by* + *-ing* to say how you do something. The *by* clause can go at the beginning or the end of the sentence.
> **By adding** a cube of plain chocolate to coffee just before using, you can restore the aroma.
>
> **Giving advice: *if* clauses**
> You can use *if* + present simple to describe a common problem, followed by an imperative or *you can* to describe what to do about it.
> **If** a lettuce becomes limp, **place** it in a saucepan with a lump of coal.
> **If** an oven dish gets burnt, **you can** put it face down on the grass all night.
> This kind of *if* clause is often called the zero conditional. For more information, see Lesson 31.

1 Write full answers to the questions in *Vocabulary and reading* activity 2 by using *to* + infinitive.

To make lettuce crisp, place it in an empty saucepan with a lump of coal.

2 Rewrite the sentences in 1 using *by* + *-ing*.

By placing lettuce in a saucepan with a lump of coal, you will make it crisp.

3 Write sentences giving advice for these problems starting with *if*...

1 Your coffee has lost its aroma.
2 You want to test if an egg is fresh.
3 You want shiny, bright teeth.
4 You get food on your tie.
5 There are flies on your windscreen.

SOUNDS

1 Circle the word in each group which has a different vowel sound.

1 wrap   ash   water   rag
2 crisp   lid   limp   ice
3 lump   cube   oven   brush
4 scrape   stain   spread   shaving

Listen and repeat the words.

2 Look at this passage and underline the /ə/ and /ɪ/ sounds.

If lettuce becomes limp, place it in an empty saucepan with a lump of coal. Put the lid on, and after a few hours it will become crisp again. Fresh coffee, once opened, soon loses its aroma. By adding a cube of plain chocolate to coffee just before using, you can restore the aroma.

Listen and check. Then read the passage aloud.

LISTENING AND WRITING

1 What would you advise for the following problems?

- tight shoes
- hiccups
- marks on wooden furniture
- chewing gum on clothes

2 Listen and find out what people advise for the problems in 1.

3 Write down the Superhints in activity 2. Use some of the vocabulary in *Vocabulary and reading* 1.

4 Think about your Superhints for three of the following:

- preparing for an exam
- learning vocabulary
- staying healthy
- cooking
- travelling long journeys
- looking good
- keeping the house clean
- making friends

5 Go round and get some Superhints from other people in the group.

6 Write a class collection of Superhints on a large piece of paper and place it where everyone can see it.

34 | *The green tourist*

First conditional

READING AND VOCABULARY

1 Work in pairs. Make a list of the advantages and disadvantages of tourism.

2 Read *Are you a green tourist?* and find out if *you* are a green tourist.

3 Work in pairs. Decide if the suggestions concern the economic, cultural or environmental effects of tourism.

4 Decide where these sentences go in the passage.

a For example, if everyone else is having a siesta, have one yourself!

b In other places, they ask for money.

c However, there is not always enough for the locals.

d As a guest the relationship between you and the host is quite different.

What words helped you make up your mind?

5 Look at the words in the box and check you know what they mean.

appreciate avoid cause consumption
cultural damage economic embarrass
energy environmental hospitable
lifestyle local photo pollution
popular public sights transport

Some words can go together.
avoid damage

Can you think of words which go with the other words?
appreciate – the countryside, the people, good food

Are you a green tourist?

Tourism around the world is so popular that in certain places, it affects and even causes damage to the sights that the tourists have come to see. It is important to think about the economic, cultural and environmental effects of being a tourist. So, before you go on holiday, here are some suggestions on how to be a green tourist.

1 **Use public transport.** If everyone uses their cars, pollution and traffic congestion will become an enormous problem.

2 **Stay in small hotels and eat local food.** It's important that the money you spend on accommodation and food remains within the local area.

3 **Travel out of season.** It's the best time to avoid crowds, and it's often cheaper as well.

4 **Think of yourself as a guest, not a tourist.** As a tourist, you're simply a source of money.

5 **Learn the local language.** If you make an effort to speak their language, you'll be able to talk to local people, and they are likely to be even more hospitable.

6 **Be careful about taking photos.** In some places, people are embarrassed when you take their photo. Find out what the local custom is.

7 **Find out about the place you're visiting.** It's very impolite to the local people if you're only there because of the weather and don't want to know anything about where you are.

8 **Use less water than at home.** In certain places, the authorities supply the big hotels with water.

9 **Use local guides.** This will create jobs and help the local economy.

10 **Adopt the local lifestyle.** If you don't appreciate being in a foreign country, why leave home in the first place?

Adapted from *The Good Tourist* by Katie Wood and Syd House

GRAMMAR

> **First conditional**
> You use the first conditional to talk about a likely situation and to describe its result. You talk about the likely situation with *if* + present simple. You describe the result with *will* or *won't*.
> *If* you **speak** the language, the locals **will be** more hospitable.
> *If* you **don't use** public transport, pollution **will become** an enormous problem.
> *If* you **don't use** less water, there **won't be** enough for the locals.

1 Match the two parts of these sentences.

1 If you like mountains,	a we'll hire a car.
2 If you don't enjoy sunbathing,	b we'll show you the sights.
3 If we have enough money,	c we'll come and visit you.
4 If we come to London,	d we'll go to the Caribbean.
5 If you stay with us in Paris,	e you'll love Britain in the summer.
6 If we want to travel around,	f you'll enjoy the Alps.

1 If you like mountains, you'll enjoy the Alps.

2 Complete the sentences with your own plans.

1 If I have enough money,	4 If there is nothing to do tonight,
2 If I have time this weekend,	5 If I work hard next week,
3 If my friends are doing nothing,	6 If I learn English,

3 Look at *Are you a green tourist?* again and rewrite each suggestion with a first conditional.

1 If you use public transport, there will be less pollution.

SOUNDS

1 🔲 Listen and repeat these phrases.

I'll I'll do that	he'll he'll call you	we'll we'll love it
you'll you'll like it	she'll she'll show you	they'll they'll learn it

2 🔲 Listen to the sentences in *Grammar* activity 1. Notice how the intonation rises in the *if* clause and falls in the main clause.

Now say the sentences aloud.

LISTENING AND SPEAKING

1 Look at the photos in this lesson. Would you like to go to any of these places? Does this kind of holiday appeal to you?

2 What are the advantages and disadvantages of the following?

– holidays in high season
– staying in hotels   – camping
– staying in the countryside
– taking your car   – flying
– travelling by public transport

3 You're going to hear two people, Max and Susie, discussing their plans to visit Nepal. Match the possible situations in the first list with the results in the second list.

1 visit Nepal in April
2 go there in August
3 go there out of season
4 fly there
5 drive overland
6 stay in hotels
7 go camping
8 take our own food
9 pay for the holiday in Britain
10 take a guide book

a travel around more easily
b meet fewer tourists
c avoid hiring a local guide
d have good weather
e spend a lot of money
f save money
g have bad weather
h get there quicker
i have a lot to carry
j use British money

🔲 Now listen and check.

4 Work in pairs. Use the situations and results in 3 to talk about Max and Susie's trip to Nepal.

If they travel out of season, they'll avoid the crowds, and it'll be cheaper.

5 Think of places in your country or around the world which attract lots of tourists? Suggest what to do about the effects of tourism.

If they stop traffic in the centre of Florence, it will be more pleasant for tourists and locals.

35 | *Lost in the Pacific*

Past perfect simple and past perfect continuous

The disappearance of Amelia Earhart during the first round-the-world flight in 1937 remains a mystery. Was she a US spy, captured by the Japanese, or did she simply crash?

Amelia Earhart was born on 24 July 1897 in Atchison, Kansas. As a child she was never really happy. But when she left home in 1920 and went to California, she learned to fly, which she loved. In October 1922, she flew to 14,000 feet, establishing a new altitude record for women. In 1928 she became the first woman to fly across the Atlantic. Then in 1929 she set a new women's speed record of over 184 miles an hour. In 1931 she married George Putnam, a publisher. Two months after they had married, Amelia became the first person to fly across the USA coast-to-coast.

In 1932 Amelia flew the Atlantic solo, and in 1934 she became the first person to cross the Pacific on a solo flight, from Hawaii to the US mainland. When she arrived, a huge crowd, newsreel cameras and a cable from President Roosevelt had been waiting for several hours to greet her. She became a magazine correspondent, a lecturer and even a fashion model. She was a woman who had achieved success in a male-dominated world.

But throughout it all, Amelia's relationship with her family was distant and she rarely visited them. There were also rumours that her marriage with Putnam had become strained.

Fame had exhausted Amelia and she decided to retire. She wanted to attempt one more record flight, around the equator.

Her first attempt in March 1937 achieved a new trans-Pacific speed record, but ended when her plane crashed in Honolulu while it was taking off.

Financially, things were not good for the Putnams, because the world flight had cost them more money than they had been expecting. But two months later, she began her second attempt on the world-record flight.

Amelia and her co-pilot disappeared without trace on the morning of 2 July 1937 somewhere in the Pacific. The accepted version of the events is that they had been running low on fuel, and had crashed into the sea and drowned. But the US Navy had been receiving SOS messages from Amelia for four days after her disappearance. It appears that the Japanese military took them to Saipan, where several people said they had seen them between August and October. There was even speculation that Amelia became involved in broadcasting radio propaganda from Tokyo. Other evidence suggests that she survived the Pacific War and returned to America, under US government protection, to live out her life in privacy. Perhaps one day the true fate of Amelia Earhart will come to light. But until then, Amelia Earhart still remains one of the world's most famous missing persons.

Life Stories, Amelia Earhart: The unsolved mystery by Randall Brink from *Marie Claire* magazine

READING

1 Read the passage and choose the best title.

1 The unsolved mystery
2 A feminist in the thirties
3 The world's most famous woman pilot

2 Number these events in the order they happened.

achieved a Pacific speed record ☐
established a new altitude record ☐
married George Putnam ☐
flew coast-to-coast across the USA ☐
disappeared in the Pacific ☐
became the first woman to fly the Atlantic ☐
set a women's speed record ☐
learned to fly ☐

Now work in pairs and check.

GRAMMAR

> **Past perfect simple**
> **You use the past perfect simple:**
> **– to talk about one action in the past which happened before another action in the past. The second action is often in the past simple.**
> *Fame **had exhausted** Amelia and she **decided** to retire.*
> **– in reported speech or thoughts.**
> *People **said** they **had seen** Amelia in Saipan.*
> **– with *after, when, because* and *until* for the first of two actions.**
> *After they **had got** married, Amelia flew across the USA.*
> **You can use two past simple tenses if you think the sequence of actions is clear.**
> *She **left** home in 1920 and **went** to California.*
> **You form the past perfect simple with *had* + past participle.**
>
> **Past perfect continuous**
> **You use the past perfect continuous when you want to focus on an action which was in progress up to or near a time in the past, rather than a completed event. You often use it with *for* and *since*.**
> *When she arrived a crowd **had been waiting for** several hours to greet her.*
> **You form the past perfect continuous with *had been* + present participle.**

1 Join the sentences in *Reading* activity 2 with *after* + past perfect for the first action, and past simple for the second action.

After she had learnt to fly, she established a new altitude record.

2 Which sentence is true? What does the false sentence mean?

a She grew tired of her fame when she disappeared.
b She had grown tired of her fame when she disappeared.

3 Choose the correct verb form.

1 Amelia *had lived/had been living* in Kansas for 23 years when she *went/had gone* to California.
2 Because she *had spent/had been spending* an unhappy childhood, she *didn't visit/hadn't been visiting* her family.
3 She *had grown/had been growing* tired of being famous when she decided to retire.
4 She *had flown/had been flying* for seventeen years when she *disappeared/had disappeared*.

4 Work in pairs and answer the questions about Amelia Earhart using *because* and the past perfect if possible.

1 Why did she maintain a distant relationship with her family?
2 Why did she decide to retire?
3 Why is it possible that they did not die on 2 July 1937?
4 What do you think happened to her?

VOCABULARY AND WRITING

1 Work in pairs. Here are some useful words from the passage. Can you remember the sentence you saw each word in. Take it in turns to make sentences similar to the original sentence. Score a point for each similar sentence, and no points if you can't remember.

> disappearance round-the-world mystery capture
> crash record coast-to-coast solo male-dominated
> rumour strained exhaust retire equator speculation
> drown survive propaganda protection privacy

2 Do you know of any other unsolved mysteries? Write a paragraph describing:

– what the mystery is
– who and what it involved
– any particular incidents

Many people claim to have seen a monster in Loch Ness.

Progress check 31–35

VOCABULARY

1 Here are some food words which English has borrowed from other languages. Are there any from your language?

chilli   pasta   casserole   cuisine   hamburger   avocado   tacos   goulash kebab   delicatessen   samovar   omelette   spaghetti   frankfurter

Do you use any of the words in your language? What words from other languages has your language borrowed? Think about:

politics – *junta, putsch, guerrilla*
music – *piano, concerto*
sport – *judo, karate*
entertainment – *rumba, samba*

2 Some words go naturally in pairs and usually in a certain order. You can say *cup and saucer, knife and fork*, but you don't usually say *saucer and cup, fork and knife*.

Here are some food words which go in pairs. What is the usual order of the words?

1	butter bread	5	milk sugar	9	cheese bread
2	fish chips	6	tea biscuits	10	salt pepper
3	jam toast	7	pears apples	11	vinegar oil
4	eggs bacon	8	cream strawberries	12	fruit vegetables

If you come across words which go in pairs, it's a good idea to write down the usual order.

3 Many words have antonyms, or opposites. It's often useful to note down new words with their antonyms.

public – private, to damage – to mend, useless – useful, faithful – unfaithful

Write down the antonyms for these words from *Reward* Intermediate.

arrival   round trip   exciting   child   lucky   number   freezing   sweet finger   buy   officer   fictional   dirty   funny   disappear   customer   laugh poverty   ceiling

4 Look at the vocabulary boxes in Lessons 31–35 again. Choose words which are useful to you and group them under headings of your choice in your *Wordbank*.

GRAMMAR

1 Complete these sentences with *in case* or *if*.

1 I'll take some food ____ I get hungry.
2 Can you buy me a loaf ____ the shop is open?
3 There may be a pool at the hotel, so pack your swimming costume ____ there is.
4 She always walked slowly ____ there was snow on the ground.
5 I'll pack my favourite soap ____ I can't get any while I'm there.
6 I'll buy a new car ____ I have enough money.

2 Reply to these questions using *when, after* and *as soon as* in turn.

1 When will the lesson finish?
2 When will you leave for home?
3 When will you sit down and relax?
4 When will you do your homework?
5 When will you have dinner?
6 When will you go to bed?

3 Make sentences with *if* and the first conditional.

1 leave now/catch your train
2 stay in bed/feel better
3 work hard/get a good job
4 eat carrots/be able to see in the dark
5 go shopping/spend a lot of money
6 ride a bike/save energy

1 *If you leave now, you'll catch your train.*

4 Rewrite the sentences in activity 3 with *by + -ing.*

1 By leaving now, you'll catch your train.

5 Join the sentences using *because* or *when* and the past perfect.

1 They left the window open. Someone broke into their house.
2 She had a busy day. She went to bed early.
3 The train already left. He arrived at the station.
4 We had a good meal. I left a large tip.
5 She lost her chequebook. She couldn't write a cheque.
6 The phone rang several times without any answer. He hung up.

1 Because they had left the window open, someone broke into their house.

SOUNDS

1 How do you pronounce the following words? Put them in five groups according to the way you pronounce the letter o: /ɒ/ , /əʊ/ , /ʌ/ , /uː/ or /ɔː/.

afford   clock   done   lose   mobile   phone
popular   one   solo   spot   torn

📼 Listen and check.

2 📼 Some words change their stress when they change their part of speech. Listen and underline the syllable which is stressed.

to suspect – a suspect   to protest – a protest
to reject – a reject   to increase – an increase
to export – an export   to record – a record
to present – a present   to produce – produce

Which syllable is stressed when the word is a verb?
Which syllable is stressed when the word is a noun?

3 Look at the following words from Lessons 31–35, and underline the stressed syllable.

consumption   economic   embarrass   hospitable
headlight   speedometer   windscreen   indicator
thunderstorm   danger   disappear   indifferent
equator   frustration   propaganda   protection

📼 Now listen and check.

SPEAKING AND LISTENING

1 Work in pairs. Read this advice for visitors to Britain. Some of the information is true and some is false. Which do you think is the false information?

When you enter a railway carriage, shake hands with everyone.
You can pick up all the fruit in a greengrocer's to choose the best.
In a pub, you go to the bar to order your drinks, and pay for them immediately.
The best way to make friends in a pub is to say, 'Can I buy everyone a drink?'
If you need help in a shop, clap your hands to call the assistant.
Don't be surprised if taxi drivers or shop assistants call you 'love' or 'darling'.
There is an excellent campsite for caravans and tents in the gardens of Buckingham Palace.
If you are stuck in a traffic jam, it's customary to sound your horn.
It's perfectly acceptable to read someone else's newspaper over their shoulder.
You always say 'have a nice day!' when you say goodbye.

2 What do you think will happen if you follow the false advice?

3 📼 Listen to John and Gertrude talking about the advice. Did you guess correctly in 1?

4 What did John say will be the results if you follow or don't follow the advice?
📼 Listen again and check.

85

Second conditional; giving advice

1 **I'm going out with a really nice guy.** He says he loves me, and he even wants us to get married. The problem is, I don't love him. I'm in love with his best friend and it's driving me crazy. I dream about him all the time and often cry myself to sleep. I've tried to find out how he feels, but he's in a difficult position too. My friends say I should stop seeing my boyfriend. But, if I can't have the boy I love, surely he is the next best thing? *Julia, 19*

2 **I've got a great girlfriend** and we've been together for about six months. It's all been quite serious but now I'm not so sure. You see, I saw my ex-girlfriend again at a party last Saturday and I think she feels we made a big mistake and that she really wants us to start going out together again. I think I feel the same. What do I do? How do I tell my present girlfriend that I think we should break up? *Barry, 21*

3 **My best friend and his girlfriend** have been together for some time, but she keeps flirting with me and I don't know if he realises or not. Should I tell him? I don't really fancy her, but I really like his ex-girlfriend. Do you think he would mind if I asked her out? He's an old friend and I don't want to upset him. *Tony, 21*

4 **I met an old boyfriend the other day** and we agreed we wanted to stay friends, although I don't want to go out with him again. We'd like to have a drink together one evening. Should we tell his present girlfriend or is it better that she doesn't know? After all, it will be completely innocent. *Sharon, 19*

5 **I have known the girl next door for a long time.** She's going out with someone who is well-known for fancying everyone in sight. The trouble is, she doesn't realise this. I happen to think he's seeing someone else, and I don't know whether to tell her or not. It's made more complicated by the fact that after all the years as childhood friends, I think I'm falling in love with her. What should I do? Should I tell her how I feel? *Steve, 20*

a Are you really sure about your ex-girlfriend's feelings for you? Don't you think you might be making another mistake? If I were you, I would talk to her a bit more before you say anything to your present girlfriend.

b You obviously like your friend very much. If he really is a good friend, tell him about his girlfriend's behaviour. He may not be pleased, but he will respect your honesty. And ask him how he feels about you asking his ex-girlfriend out.

the problem **page**

c I really think both of you should tell his girlfriend. If she found out from someone else she would be very upset. Put yourself in her position, and decide how you would feel.

d If she doesn't feel for you the way you feel for her, she won't be pleased to be told about her boyfriend's behaviour. Find out if she fancies you. If she does, then you can tell her the news about her present boyfriend.

e I think you really should work out your feelings for your boyfriend. If you're attracted to his best friend, it doesn't look very promising. If I were you, I'd take a break from both boyfriend and his best friend and see how you feel in three months' time.

READING AND VOCABULARY

1 Work in pairs. Here are some words and phrases you can use to describe relationships. Look at the verb and verb phrases. Put them in the order in which you can use them to describe the stages of a relationship.

> acquaintance friend girlfriend fancy someone
> fall in love with someone partner ask somebody out
> meet someone go out with fall out with someone
> lie about something get engaged to someone
> do something behind someone's back boyfriend
> get married to someone get involved with someone
> break up feel jealous fiance(e) get to know someone

meet someone, get to know someone...

Now look at the nouns. Which stages of the relationship do they go with?

meet someone: acquaintance

2 Read the problem page letters and match them with the replies.

3 The people in the problem page letters are all connected. Can you say who is who? Use the words and expressions in the box above to say what the relationship between each person is.

GRAMMAR

> **Second conditional**
> You use the second conditional to talk about an imaginary or unlikely situation in the present or future and to describe its result. You talk about the imaginary or unlikely situation with *if* + past simple. You describe the result with *would* or *wouldn't*.
> *If I left my fiancée, she would be very upset.*
> *If they got married, it wouldn't last long.*
> *If he wasn't so attractive, I wouldn't worry so much.*
>
> **Giving advice**
> You can use the following expressions to give advice.
> *If I were you, I wouldn't let it spoil the relationship.*
> *I think you should/ought to talk about your feelings.*
> *In my opinion, she should/ought to wait until she has completed her studies.*

1 Write sentences saying what you would do if:

1 a friend lied to you
2 you fell out with a friend
3 you felt unhappy about a relationship
4 your parents didn't approve of your friends
5 your partner wanted to stay at home all the time

2 Work in pairs. Talk about your advice to the people mentioned in the letters.

SOUNDS

Look at the sentences below. Underline the words you think the speaker will link.

1 If I were you, I'd invite him in.
2 In my opinion, you ought to eat it all up.
3 If I were you, I wouldn't even answer.
4 In my opinion, you ought to wear a suit.
5 If I were you, I'd turn on the heating.

Listen and check. Say the sentences aloud.

LISTENING AND WRITING

1 You are going to hear some people talking about one of the following situations. First, read them and decide what you would do.

1 A friend wants to come and stay with you. She is an old friend, but smokes all the time, has a shower three times a day, throws away the newspaper before you've read it and uses the place like a hotel. She wants to stay a week.

2 It's ten o'clock on a Monday evening. You're quite tired and you need to do some work before you go to bed. There is a knock at the door. A close friend is standing there. He looks upset and needs someone to talk to.

2 Listen and decide which situation the speakers are talking about.

3 Choose one of these situations and write a letter giving your advice.

Past modal verbs (1): *should have*

VOCABULARY AND READING

1 You're going to read an extract from a guide to Hong Kong. Put the words in the box under the following headings: *where to stay, what to do, what to buy, when to go, what to eat, what to wear, where to go.* (Some words can go under more than one heading.)

peak season summer autumn humid view spectacular junk harbour accommodation bikini miniskirt shorts sandals lobby dish steam bamboo basket shopping mall silk porcelain relax

2 Read *Hong Kong Factfile* and match the paragraphs with the headings in 1.

3 Answer the questions about words which may be new to you.

scarce – Is the ideal time likely to be when there are many or few tourists? (paragraph 1)

dirt-cheap – Is this likely to be very cheap or very expensive? (paragraph 3)

hassle – Is it difficult or easy to find a room? (paragraph 3)

stuff – Do these people offer you the advertisements, or force you to take them? (paragraph 6)

4 Is this information about visiting Hong Kong true or false?

1 The weather is uncomfortably hot in summer.
2 Most people who visit Hong Kong go to Victoria Peak.
3 There are plenty of cheap hotels in Hong Kong.
4 You can wear anything you want.
5 *Dim sum* is too much to eat on your own.
6 You can buy anything you want in Hong Kong.
7 It's not a relaxing place.

Hong Kong Factfile

1 Anytime is OK but summer is the peak season, which means higher airfares and sometimes a shortage of hotel rooms. The weather is hot and humid as well. The Chinese New Year is also very busy and so is Easter. The ideal time is autumn when the weather is at its best and tourists relatively scarce.

2 If you haven't been to the Peak, as Victoria Peak is usually called, then you haven't been to Hong Kong. Every visitor makes the trip and for good reason – the view is one of the most spectacular in the world. Aberdeen is also a top tourist attraction, where nearly six thousand people live or work on junks anchored in the harbour. Also moored in the harbour are three palace-like floating restaurants, sightseeing attractions in themselves.

3 If you're looking for dirt-cheap accommodation, Hong Kong is no paradise. High land prices mean high prices for tiny cramped rooms. Even though new hotels are constantly being built, demand always seems to exceed supply. However, you can usually find a room without too much hassle, although this can be difficult during peak holiday times like Chinese New Year and Easter.

4 Hong Kong is a cosmopolitan and fashion-conscious city. The Chinese generally judge a person by their clothing far more than a Westerner would. Revealing clothing is OK – shorts, miniskirts and bikinis (at the beach only)

are common. Although Hong Kongers are very tolerant when it comes to dress, there is one exception – flip-flop sandals (thongs). Flip-flops are OK to wear in your hotel room or maybe the corridor, but not in its lobby and most definitely not outdoors (except around a swimming pool or beach).

5 One of the big rewards of coming to Hong Kong is the opportunity to try *dim sum*. *Dim sum* is dumplings which are usually steamed in a small bamboo basket. It is a uniquely Cantonese dish served only for breakfast or lunch, but never dinner. Eating *dim sum* is something you should do in a group. It consists of many separate dishes which are meant to be shared. Typically, each basket contains four identical pieces.

6 Hong Kong is a shopper's paradise. If you can't find it in Hong Kong, it probably doesn't exist. In Tsimshatsui, people are constantly trying to stuff advertisements into your hands. Try the shopping malls such as the City Plaza in Hong Kong island or the Tuen Mun Town Plaza in the New Territories. Best buys are computer equipment, electrical appliances, cameras, clothes, especially things in silk and men's suits and porcelain.

7 Go shopping, eat out, go sightseeing, go to the beach, take a trip to the New Territories, take a ferry to the islands, relax – what do you mean relax? There's far too much to do.

Adapted from *Hong Kong – a travel survival kit* by Robert Storey

LISTENING

1 🔲 Listen to Jill talking about a visit she made to her brother who lives in Hong Kong. Which of the things mentioned in *Hong Kong Factfile* did she do?

2 Work in pairs and check your answers to 1. Which things didn't Jill do?

🔲 Listen again and check.

GRAMMAR

> **Past modal verbs (1):** *should have*
>
> **You use *should have* and *shouldn't have* to describe actions in the past which were wrong, or which you now regret.**
> She **should have** gone to Hong Kong in the autumn.
> She **shouldn't have** gone in the summer.
> **You often use it to criticise someone.**
> She **shouldn't have** worn flip-flops in the restaurant.
> **Ought/oughtn't to have has a similar meaning to should/shouldn't have.**

1 Write sentences with *should* or *shouldn't have.*

1 It was wrong of me to walk. It's quicker by car.
2 Are you still waiting for her? There's no need.
3 He has eaten too much and now he feels ill.
4 They spent too long in the sun and now they've got sunburn.
5 It's a pity you weren't here last week. There was a festival of music.
6 My bags are so heavy with presents to take home. I can't lift them.

1 I shouldn't have walked. I should have taken the car.

2 Work in pairs. Say what Jill *should* or *shouldn't have* done. Use your answers to *Listening* activities 1 and 2 to help you.

SPEAKING

1 Have you ever done anything which was wrong and which you now regret? Think about:

– someone you met   – somewhere you went
– something you bought   – something you ate
– something you said   – something you did

2 Work in groups of three or four and talk about what you did wrong. Do the others think it was a serious mistake?

Now you see me, now you don't

Past modal verbs (2): *may have, might have, could have, must have, can't have*

READING AND SPEAKING

1 Work in pairs. In this lesson there are some true stories about people appearing or disappearing. Look at these questions about the first story and predict what the answers might be.

1 Where was the old Spanish woman looking after her grandchild?
2 Why did the grandchild cry out?
3 Where was the strange, sad face?
4 What appeared three weeks later?
5 What did they hear at the same time?
6 What happened in the end?

2 Read the story and put the paragraphs in the right order. The ending is missing.

a For what had frightened the child was a strange, sad face staring up from the faded pink tiles of the kitchen floor. When the woman had recovered from the shock, she tried to rub away the face. But the eyes only opened wider, making the expression of the face even sadder.

b When the kitchen was locked and sealed, four more faces appeared in another part of the house and microphones set up by investigators recorded sounds the ear could not hear – moans and voices which were speaking in a strange language. The faces and sounds both disappeared, as mysteriously as they had arrived, leaving no clue as to what they were or why they had come.

c After the incident, in August 1971, the old woman sent for the owner of the house, who lived in Belmez near Cordoba. He removed the tiled floor and replaced it with concrete. But three weeks later another face appeared, its features even clearer. Now faces appeared all over the kitchen floor - first one, then another, then a whole group.

d An old Spanish woman was looking after her grandchild in the kitchen of her tiny village home when the youngster suddenly cried out. The grandmother turned round - and got the shock of her life.

Now work in pairs and check your answers to 1.

GRAMMAR

> **Past modal verbs (2):** *may have, might have, could have, must have, can't have*
>
> **You use *may have, might have* and *could have* to talk about something that possibly happened in the past.**
> *The child may have imagined it.*
> *The old woman might have seen a ghost.*
> *It could have been a trick.*
>
> **You use *must have* and *can't have* to talk about something that probably or certainly happened in the past.**
> *It must have been frightening.*
> *It can't have been a trick.*
>
> **You can use a verb in its continuous form with past modals.**
> *The faces must have been trying to say something.*
> *The voices can't have been speaking in Spanish.*

1 Look at these sentences and explain the difference between them. Which mean the same?

1 It might have been a ghost.
2 It must have been a ghost.
3 It could have been a ghost.
4 It can't have been a ghost.

2 Write sentences speculating what might have or could have happened.

1 I can't find my wallet. I wonder where it is.
2 The car won't start.
3 She should be here by now.
4 I've rung him several times but there's no answer.
5 The door is closed but it isn't locked.
6 The car is in the garage but his bike has gone.

1 I might have left it at home.

3 Write sentences drawing conclusions about what *must have* or *can't have* happened.

1 She looks pale and very thin.
2 He's been yawning all morning.
3 She looks very upset.
4 When the train arrived, they weren't on it.
5 I'm sure he knocked over the vase by accident.
6 The pavement is very wet.

1 She must have been ill.

4 Work in pairs. Talk about what might have happened in the story in *Reading and speaking 2*. Use *might have, must have, can't have*.

Now turn to Communication activity 16 on page 102 and find out what had happened.

READING

Work in pairs. You're going to read some more stories about people who disappeared.

Student A: Turn to Communication activity 4 on page 99.
Student B: Turn to Communication activity 12 on page 101.

SPEAKING AND VOCABULARY

1 Work in pairs. Tell each other the stories you read.

2 Work in pairs and talk about possible explanations to the stories. Use *could have, might have, must have* and *can't have*.

3 Turn to Communication activity 17 on page 102 to find the explanations to the stories.

4 Here are some new words from this lesson. Can you remember which stories they appeared in?

> assume   bandage   chatter   clue   companion   cover   disappear   disaster
> explore   face   floor   lock   manhole   moan   mysteriously   pay attention   plague
> seal   side street   tiles   youngster

'Bandage' must have been in the first story.

5 Work in groups of two or three. Do you know of any other strange stories like the ones in this lesson? Think about:

ghosts   crimes   ships lost at sea   strange things in the sky   wild animals
fortune telling   voices from the past   reincarnation   curses

Do any of your stories have simple explanations?

39 | *Making the grade*

Expressing wishes and regrets

John di Salvador and Frances Graham both live in Birmingham and have just finished their state education after sixteen years in schools and colleges. But John lives in Birmingham, Alabama, in the United States of America, and Frances lives in Birmingham, England. So how do the two educational systems compare?

VOCABULARY AND LISTENING

1 The words in the box below can be used to describe the British or the American education systems. Do you know which words are British and which are American? Put GB or US by each word.

> kindergarten grade grade school elementary school
> primary school junior high school middle school
> secondary school high school comprehensive school
> public school boarding school

2 📼 Listen to John and Frances talking about the education systems in their countries. Did you guess which words were British or American in 1?

3 Put these subjects in the box below under three headings: *primary school, secondary school, university.*

> art geography history music physical education
> science and technology maths languages chemistry
> physics car repair economics computer studies
> reading writing law medicine philosophy typing

Are there any subjects which you can't study at school or university, but which you think you should?

4 📼 Work in pairs.

Student A: Turn to Communication activity 5 on page 99.

Student B: Turn to Communication activity 13 on page 101.

5 Work together and complete the chart below.

	Britain	USA
Starting age		
Leaving age		
Examinations: type and age		
Corporal punishment		
Homework		
Classroom participation		
Payment of university fees		
Entrance requirements		

GRAMMAR

> **Expressing wishes and regrets**
> You can express wishes with **I wish** + past simple.
> *I **wish** I **was** a student.*
> **(I'm not a student, but I would like to be.)**
> You can express regrets about the past with **I wish** + past perfect.
> *I wish I **had worked** harder.*
> **(I didn't work hard, and I regret that.)**
> You can usually use **If only** instead of **I wish.**
> **If only** *I was a student.* **If only** *I had worked harder.*

1 Explain the difference between these sentences.

a I wish I spoke English.
b I wish I had spoken English.

2 Write explanations of the meaning of these sentences.

1 I wish I could type.
2 I wish I had learnt to play the piano.
3 I wish I lived in America.
4 I wish I didn't smoke.
5 I wish I had stayed in touch with my school friends.
6 I wish I could travel round the world.

1 I would like to be able to type.

3 Which sentences in 2 show a wish for the future and which ones show a regret about the past? Rewrite them using *If only*.

If only I could type.

4 Write sentences describing what Frances and John regret.

🔲 Listen again and check.

SOUNDS

1 🔲 Listen and tick the sentence you hear.

I wish I worked harder.	I wish I'd worked harder.
If only I walked to work.	If only I'd walked to work.
I wish I kept my temper.	I wish I'd kept my temper.
If only she paid me.	If only she'd paid me.
I wish I had a dog.	I wish I'd had a dog.
If only he liked me.	If only he'd liked me.

Say the sentences aloud.

2 🔲 Listen and check your answers to *Grammar* activity 3. Notice how the intonation rises and falls in sentences with *if only*.

Now say the sentences aloud.

WRITING AND SPEAKING

1 Write down four or five important events in your educational career.

When I was ten, I started to learn English.

Is there anything you wish you had done differently?

2 Now work in pairs. Tell each other about these important events. Ask and say what you regret about the past and what you hope for in the future.

I wish I had studied English at school.

3 Write a description of the educational system in your country and compare it with Britain or the USA. Use the words in *Vocabulary and listening* activities 1 and 3 and the chart in activity 5 to help you. Use *like* + country to introduce things which are the same in Britain or the USA.

Like *Britain, university fees are free.*

Use *but unlike* + country to introduce things which are different.

But unlike *Britain, we don't receive any money for accommodation.*

Third conditional

He first noticed (1) ___ on a Tuesday evening, on his way from the station. The man was tall and thin, with a look about him that told Ray Bankcroft (2) ___ It wasn't anything Ray could put his finger on, the fellow just looked English.

That was all there was to their first encounter, and the second meeting passed just as casually, (3) ___ The fellow was living around Pelham some place, maybe in that new apartment house in the next block.

But it was the following week that Ray began (4) ___ The tall Englishman rode down to New York with Ray on the 8:09 train, and he was eating a few tables away at Howard Johnson's one noon. But (5) ___ Ray told himself, where you sometimes ran into the same person every day for a week.

It was on the weekend, when Ray and his wife travelled up to Stamford for a picnic, that he became convinced (6) ___ For there, fifty miles from home, the tall stranger came striding across the rolling hills, pausing now and then to take in the beauty of the place. 'Linda,' Ray remarked to his wife, 'there's that fellow again!'

'What fellow, Ray?'

'That Englishman from our neighbourhood. (7) ___'

'Oh, is that him?' Linda Bankcroft frowned through the tinted lenses of her sunglasses. '(8) ___'

'Well, he must be living in that new apartment in the next block. I'd like to know what he's doing up here, though. Do you think he could be following me?'

'Oh, Ray, don't be silly,' Linda laughed. 'Why would anyone want to follow you? (9) ___'

'I don't know, but it's certainly (10) ___'

It certainly was odd.

GRAMMAR

> **Third conditional**
>
> You use the third conditional to talk about an imaginary or unlikely situation in the past and to describe its result. You talk about the imaginary or unlikely situation with *if* + past perfect. You describe the result with *would/wouldn't have* + participle.
> *If I'd seen him, I would have ignored him.*
> *If he'd spoken to me, I wouldn't have said anything.*
> You separate the two clauses with a comma. You can also use *may have, might have* and *could have* if the result is not certain.
> *If he had followed me, I might have told him to stop.*

1 Say what would have happened if things had been different.

1 Giles/catch the plane/not spend his holiday at home
2 Brenda/not forget her chequebook/buy us lunch
3 Jane/stay at home/not feel so ill
4 Andrea/listen to the radio/hear the bad news
5 the police arrived quickly/catch the burglar
6 Karim/speak English/make himself understood

1 If Giles had caught the plane, he wouldn't have spent his holiday at home.

2 Think of four or five memorable events in your life.

We had a baby.

Now work in pairs. Tell each other about these important events. Ask and say what you would have done if the events hadn't happened.

What would you have done if you hadn't had a baby?
We would have got a cat.

READING AND LISTENING

1 You're going to read and listen to a story called *The man who was everywhere* by an American writer, Edward D Hoch. Read part 1 and guess where these extracts go in the story.

a the new man in the neighbourhood
b And to a picnic?
c the Englishman was following him.
d he was English.
e to notice him everywhere.
f odd the way he keeps turning up.
g that was the way things were in New York,
h I don't remember ever seeing him before.
i Friday evening at the station.
j The one I was telling you I see everywhere.

2 🔲 Listen to part 1 of the story and check your answers to activity 1.

3 What do you think is going to happen next? Try to imagine the answers to the questions before you listen to part 2 of the story.

1 Did the Englishman appear again?
2 What did Ray do one night on his way home?
3 What did the Englishman do?
4 How often did Ray see him now?
5 How did he feel about it?
6 What did he threaten to do?

🔲 Now listen to part 2 of the story and check your answers.

4 Work in pairs. What would you have said or done?

5 In part 3, decide who does the following actions, the Englishman or Ray. Where does their final meeting take place?

ran out of cigarettes	paused, out of breath
saw him waiting	followed him down the
was beckoning him to follow	railroad
ran on, faster and faster	turned and walked away
called out, 'Come back here!'	heard the Express train

6 Turn to Communication activity 18 on page 102 and read part 3 of the story and check your answers to 5. What happened?

7 Work in groups of two or three. This is not the end of the story. What do you think happened?

🔲 Now listen to part 4 of the story and find out.

8 The story is very tense. What aspects of the writer's style contribute to the tension?

VOCABULARY AND WRITING

1 Here are some useful words from the story. Check you know what they mean.

> neighbourhood fellow meeting casually noon picnic
> frown follow mysterious oblivious puzzled elevator
> curse threaten crazy grab beat drugstore tracks
> whistle murder skill

2 Write the story from the point of view of Ray's wife or of the Englishman.

Progress check **36–40**

VOCABULARY

1 Sometimes it's difficult to understand what an idiomatic expression means. Here are some examples:

We're on the same wavelength means *we understand each other.*
It was a storm in a teacup means *it looked like a serious dispute, but in fact, it wasn't.*

It doesn't always sound very natural for language learners to use these kinds of expressions, but it is important to be able to find out what they mean. To do this, you need to decide which is the most important word in an expression, and then look it up in a dictionary. Remember that idiomatic expressions are like vocabulary items, and need to be learnt as complete phrases.

The following idiomatic expressions are all to do with parts of the body. Use a dictionary to find out what they mean.

She's all fingers and thumbs.
We don't see eye to eye.
He's very nosy.
I can't make head nor tail of it.
I've gone and put my foot in it.
It cost me an arm and a leg.
I'm all ears.

Rewrite each expression more simply.

2 There are many idiomatic expressions with *like* and *as*.

to smoke like a chimney   *as quiet as a mouse*   *as brave as a lion*

Match the two parts of the phrases below.

to work		a fish	as drunk		a mule
to sleep	like	a horse	as thick	as	a pig
to eat		a log	as stubborn		two short planks
to swim		a dog	as greedy		a lord

Can you think of people or things you could use these expressions to describe?

3 Look at the vocabulary boxes in Lessons 31–35 again. Choose words which are useful to you and group them under headings of your choice in your *Wordbank*.

GRAMMAR

1 Say what would happen if things were different.

1 Paul/like sunshine/go to Spain
2 Jenny/work harder/get a better job
3 You/speak more slowly/be easier to understand
4 We/take the car/get there sooner
5 Graham/speak French/live in Paris

1 If Paul liked sunshine, he'd go to Spain.

2 Write your advice to these people. Use *If I were you, I'd...*, *I think you should/ought to...*, and *In my opinion, you should/ought to...*

1 'I've got a bad cold.'
2 'I get angry very quickly.'
3 'I'm always losing my umbrella.'
4 'I'd like to go to America.'
5 'I need some fresh air.'

1 If I were you, I'd go to bed.

3 Write sentences with *should have*. Use the words in brackets.

1 I took the coach and got stuck in a traffic jam. (take the train)
2 It's so hot and I've only packed my winter clothes. (bring summer clothes)
3 You've arrived very late. (leave earlier)
4 It was her father's birthday yesterday and she forgot to send a card. (ring him)
5 The hotel where he stayed was too expensive. (look for a cheaper one)

1 I should have taken the train.

4 Rewrite the sentences in 3 using *shouldn't have*.

1 I shouldn't have taken the coach.

5 Rewrite sentences saying what *might have* or *could have* happened.

1 I wonder if the plane was delayed.
2 Maybe she missed it.
3 Perhaps they lost her baggage.
4 I wonder if she came out of another exit.
5 Maybe we didn't see her as she came out.

1 It might have been delayed.

6 Write sentences saying what *must have* or *can't have* happened.

1 She didn't get my letter.
2 There's no one at the bus stop.
3 He didn't look up when I spoke to him.
4 His hair looks nice and clean.
5 That coat cost her most of her savings.
6 That child is holding its head and crying.

1 It must have got lost in the post.

7 Rewrite these sentences using *I wish*.

1 I'd love to have a new car.
2 It's a pity I didn't keep my tickets with my passport.
3 I hate it when I don't have enough room.
4 I'm sorry I failed my driving test.
5 If I had his address, we could visit him.

1 I wish I had a new car.

8 Rewrite the sentences you wrote in 7 using *If only*.

1 If only I had a new car.

9 Rewrite these sentences describing the situation in the past and its probable result with *if* and the past perfect, and the third conditional.

1 I took a taxi. I arrived in time.
2 She spoke to him in French. They got on well.
3 He apologised. I didn't lose my temper with him.
4 We didn't have any money. We didn't go to the theatre.
5 He bought a bicycle. He got fit.
6 It rained all week. We had a dreadful holiday.

1 If I hadn't taken a taxi, I wouldn't have arrived in time.

SOUNDS

1 How do you pronounce the following words? Put them in five groups according to the way you pronounce the letter u: /juː/ /uː/ /ʌ/ /ɜː/ /ʊə/

computer junior funny nurse surgeon cure rural butter usual student sun jury

Listen and check.

2 Underline the silent letters in these words.

highlight throughout known bought Christian wheel ironing eight psychiatry

Listen and say the words aloud.

3 Underline the words you think the speaker will stress.

1 If I'd done it, I'd have told you.
2 If only he was an accountant.
3 I wish I hadn't eaten so much.
4 She should have asked you first.
5 If I were you, I'd do it myself.
6 He must have opened it.

Now listen and check.

READING AND SPEAKING

1 The passage below consists of two stories which have been mixed up. Choose one of them and write it out in full.

A man was saving for a new car. It was to be a surprise, and he didn't tell his wife that he had already saved £500 and hidden it in a pile of old clothes. A man had worn a hearing aid for 20 years, but it had never seemed to help him hear better. He was out when the dustmen called and his wife gave them the old clothes. When he went to hospital for a routine check-up, he was told he was wearing it in the wrong ear. When he discovered the mistake he hired a mechanical digger to search the rubbish dump. He said, 'They must have made a mistake when they first gave it to me. I always thought it was useless.' After two days search he gave up and started saving again.

2 Choose two or three points in the story and write a sentence describing

– what should or shouldn't have happened
– what would have happened if things had been different
– what the man wished he had or hadn't done.

Communication activities

1 *Lesson 3*
Reading, activity 3

Student A: Read the following passage and find out if it mentions anything about:

preparations   dress   wedding ceremony   presents
reception

In the Arab world, the groom's family visits the bride's family and asks for her hand in marriage. When the bride's family agrees, they drink coffee and talk about the other arrangements. There are two ceremonies. Before meeting his bride, the groom attends the Moslem ceremony. The bride does not attend this ceremony but allows a sister or a friend to make the marriage promises on her behalf. Then there are two parties, one for the men and one for the women, which can last for days. After celebrating their marriage apart for several days, the couple finally get together. After arriving at the wedding reception, the bride, dressed in white, and the groom sit on a small stage and enjoy the celebration. During this part of the wedding, they receive and open the wedding presents. The presents people give are often jewellery and gold. The celebration lasts for several hours. There is usually a dinner and supper with relatives and close friends of the couple.

Now turn back to page 6.

2 *Progress check 1–5*
Speaking and writing, activity 1

3 *Lesson 6*
Reading, activity 3

Student A

Josephine Wilson

When I first met Nguyen, it was probably nine or ten weeks after he had arrived in Los Angeles, and he was feeling very lonely. He could speak very little English and he was in culture shock. He was working at the local greengrocer's and was living with a friend of mine, Cathy Kelly, in a big house with some other people, who needed housing. I met him one day when she was having a party to help people like him get to know others in the neighborhood. I liked him immediately but it was difficult to have a conversation with him because he was still learning English. He didn't have many clothes and I remember one of the first things I did was to take him shopping to buy something to wear.

His English improved very quickly. It was interesting to see how he was settling into Western life. He was coping with it all very well, even though it was very difficult for him. He comes from Ho Chi Minh city, and having been there myself, I know there couldn't be a greater contrast to Los Angeles.

At my age I don't make friends easily but I count him among my closest friends. I think he is one of the most honest people I have ever met. I'm quite surprised he likes me and wants to spend such a lot of time with an old lady. But he's kind and considerate. We are natural friends. I'm very easy in his company and I love to hear his stories about his life as a child. He comes from a culture that is so alien to ours, and yet we understand each other. And I love having a young person around, the excitement and optimism is really heartening.

Now turn back to page 15.

4 *Lesson 38*
Reading

Student A: Read the story and guess the missing words and phrases.

In 1889, an English woman and her daughter, (1) ___ , checked into one of the most lavish hotels. (2) ___ . The daughter wanted to take in the sights and sounds of the city immediately, but her mother, (3) ___ , wanted to sleep. The girl went out alone, (4) ___ , and saw the Eiffel Tower. Six hours later she returned to her mother's room, only to find it empty, and no sign of her mother ever having been there. When she checked with the front desk, they insisted that they had never seen her (5) ___ . The mother had disappeared. The desperate girl searched for weeks (6) ___ . She died several years later in a mental hospital, having gone mad (7) ___ .

Here are the missing phrases. Decide where they go in the story.

a Each had her own room
b ...on a visit to the Great Paris Exhibition...
c ...strolled down the Champs Elysées...
d ...because of the loss of her mother
e ...tired after the trip...
f ...before finally returning to England
g ...or her mother

5 *Lesson 39*
Vocabulary and listening, activity 4

Student A: 🔊 Listen to Frances talking about the education system in Britain.

Find out:

- when you can start school
- when you can leave school
- when and what exams you take
- if there is corporal punishment
- if there is homework
- who pays for a university education
- how you get into university

When the recording stops, turn back to page 93.

6 *Lesson 8*
Reading and writing, activity 2

Read the last part of Claudia's story and complete or correct the sentences you wrote in *Reading and writing*, activity 1.

I then took the train to St Petersburg. No one was waiting for me at the station to meet me, although the theatre I was visiting was going to send someone. I later learnt that the director didn't recognise me because I was wearing a fur hat and looked like a Russian. So I took a taxi to the theatre. Before I could find a hotel one of the actresses, Natasha, invited me to stay with her. She lived in a communal flat which she was sharing with six other people. As we were walking to her flat, she apologised constantly, while I tried to assure her that it was an adventure for me. Natasha couldn't feed me on her salary because she was only earning about £3.00 a month, so I tried to find food during the day. I bought food on the streets when I found it. Old women stood in the bitter cold and sold pasties which they kept warm in large pots with steaming cloths sealing them against the falling snow. I never knew what was going to be inside them and ate cabbage pasties for breakfast until I discovered a woman in front of the Winter Palace selling apple turnovers. They were the best I had ever tasted, and by the end of the week she had two apple turnovers ready when she saw me approaching. Other than beetroot and bananas I saw no vegetables or fruit all week.

But I had a very real experience of what life is really like in Russia.

Adapted from *Down and Out in Moscow* by Claudia Wolgar, *The Sunday Times magazine*

Now turn back to page 19.

7 *Lesson 12*
Writing and speaking, activity 1

Write down as much of the news item as possible. Try to get it correct word-for-word. When you have written as much as you can, compare your version with other people in the class. Try to reconstruct the news item as accurately as possible.

8 *Lesson 17*
Vocabulary and writing, activity 3

As directed by Sefton's will, the picture Aphrodite was exhibited at the Royal Academy. The suicide made a good deal of talk at the time and a special attendant was necessary to regulate the crowds. Sefton had been found in his studio many hours after his death; and he had scrawled on a blank canvas, 'I have finished it but I can't stand any more.'

9 *Lesson 25*
Reading and vocabulary, activity 1

Part 2

The car stopped, they rang the bell, the door opened, and Rosemary drew the girl into the hall. Warmth, softness, light, a sweet scent, all these familiar things she never even thought about.

The girl seemed dazed.

'Come and sit down,' Rosemary cried, as she walked into the drawing room. 'Sit in this comfortable chair. You mustn't be frightened. Come and get warm. Won't you take off your hat?'

The girl remained standing, but Rosemary gently pulled her towards the chair. The girl had moved only a few paces when she whispered, 'I'm very sorry, madam, but I'm going to faint.'

Rosemary guided her into the chair, crying, 'How thoughtless I am! Tea at once! And some brandy immediately!' She rushed to the bell to call a servant.

'No, I don't want brandy. It's a cup of tea I want, madam.' And she burst into tears.

'Don't cry, poor little thing,' Rosemary said, sympathetically.

Now at last the girl forgot to be shy and said, 'I can't go on any longer like this.'

'You won't have to. I'll look after you. Don't cry any more. I'll arrange something, I promise.'

The girl stopped crying just in time for tea.

Some little while later, the door handle turned. 'Rosemary, may I come in?' It was Philip.

'Of course. This is my friend, Miss -'

'Smith, madam,' said the girl, who was strangely still and unafraid.

Philip looked at the girl, at her hands and boots and then at Rosemary. 'I wanted you to come into the library for a moment. Will Miss Smith excuse us?'

Rosemary answered for her: 'Of course she will.' And they went out of the room together.

Philip asked, 'Explain. Who is she?'

Rosemary laughed and said, 'I picked her up in Curzon Street. She asked me for the price of a cup of tea and I brought her home with me. I'm going to be nice to her. Look after her.'

'My darling girl,' said Philip, 'you're quite mad. She's so astonishingly pretty.'

'Pretty?' Rosemary was so surprised. 'Do you think so?'

'She's absolutely lovely. Look again. I think you're making a mistake. But let me know if Miss Smith is going to have dinner with us.'

Rosemary did not return directly to Miss Smith but went to her writing room and sat down at her desk. Pretty! Lovely! She opened a drawer and took out five £1 notes. She paused, then put two back and, holding three in her hand, went back to the girl in the drawing room.

Half an hour later, Philip was still in the library, when Rosemary came in.

'I... only wanted to tell you Miss Smith insisted on going, so I gave the poor little thing some money.'

Rosemary had just done her hair, darkened her eyes and put on her pearls. She touched Philip's cheeks.

'I saw a fascinating little box today. It costs twenty-eight guineas. May I have it?'

'You may,' Philip said. But that was not really what Rosemary wanted to say.

'Philip,' she whispered, 'Am I pretty?'

10 *Lesson 3*
Reading, activity 3

Student B: Read the following passage and find out if it mentions anything about:

preparations   dress   wedding ceremony
presents   reception

While a Moslem wedding has at least two ceremonies, the traditional Chinese Taoist marriage has three. Before the man proposes to the woman, his family asks the professional matchmaker to send a present from them to the bride's family. Before agreeing to the marriage, the bride's family must accept the present. Then it is time for the second stage which, like the Hindu custom, is the checking of the horoscopes. After the matchmaker has made sure that the signs are good, the two families ask the gods for their help. Before starting the celebration, the groom's family pays the bride's family for losing a daughter. Then the wedding party begins. Both the bride and the groom are dressed in silk. The groom gives ceremonial gifts of pork, chickens, candles and clothing to the bride's family. From the guests at the reception, the couple receive red packets (*hong bao*) containing gold, jewellery or money. The reception is often a lunch or dinner of fifteen courses, with entertainment by a singer and a band.

Now turn back to page 6.

11 Lesson 6
Reading, activity 3

Student B

Nguyen Van Tuan

The first time I remember seeing Josephine was when I was working in a store. She was taking her dog for a walk and was doing her shopping when she came into the store to buy some dog food. I remember being very impressed by her politeness and friendliness to the store owner. At the time I was staying in a house owned by Cathy Kelly, another amazing lady who has this huge house which she keeps open to anyone who needs a bed for the night. There were lots of us in the house, which was good for me as I was feeling particularly lonely and was suffering from culture shock. I still do, to a lesser extent. I finally met Josephine when Cathy introduced us at her house one day. Josephine was beautiful and kind to me when we finally spoke.

I had left Vietnam a few weeks before and to cut a long story short, I ended up in Los Angeles. It was very strange, so different from what I knew back home. There was so much to do, so many opportunities. Josephine took me to concerts and political meetings and sometimes to the theatre although I understood very little. We saw quite a lot of each other in those first few months. At one point we were seeing each other every day because she was teaching me English.

A few weeks ago, I told her that I wanted to go to college and she's trying to find out if it's possible. She says she will pay for my education. I am surprised that her family let her live alone. In my country, the whole family lives in the same house, old people and young. But if she didn't live alone, I guess I wouldn't see her so often.

I place great value on the friends Josephine has helped me make and on my friendship with her.

Now turn back to page 15.

12 Lesson 38
Reading

Student B: Read the story and guess the missing words and phrases.

An English journalist was walking (1) ___ in Torremolinos with a friend, (2) ___ but not paying much attention to what the other was saying. At one stage the friend turned round to say something to the journalist and was astonished (3) ___ . He assumed his companion must have disappeared into a side street (4) ___ , so he returned to their hotel, expecting the journalist to turn up (5) ___ . Four hours later, (6) ___ when the friend was thinking about contacting the police, the journalist turned up at the hotel, (7) ___ .

Here are the missing phrases. Decide where they go in the story.

a ...in time for dinner...
b ...along a small street...
c ...with a bandage around his head
d ...to explore something of interest...
e ...long after the meal was over and...
f ...chattering happily...
g ...to find that he had disappeared

13 Lesson 39
Vocabulary and listening, activity 4

Student B: [cassette] Listen to John talking about the education system in the USA.

Find out:

– when you can start school
– when you can leave school
– when and what exams you take
– if there is corporal punishment
– if there is homework
– who pays for a university education
– how you get into university

When the recording stops, turn back to page 93.

14 *Lesson 11*
Reading

24 – 36 points.

You get on with things. But be careful! You may be a bit too efficient for most people.

12 – 23 points.

You tend to put things off but you recognise that you're not perfect.

0 – 11 points.

You put things off until the last moment. Never mind! Most people are like you.

15 *Lesson 16*
Vocabulary, activity 2

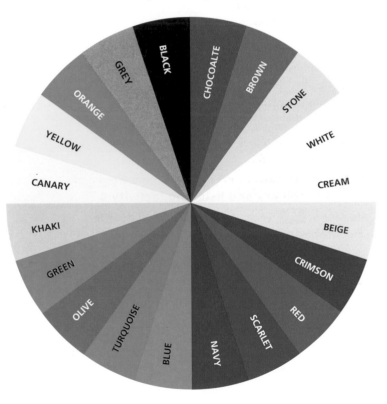

16 *Lesson 38*
Grammar, activity 4

They were very puzzled by what they had seen, so they called the local authorities. They were told that the house was built on the site of a medieval cemetery.

17 *Lesson 38*
Speaking and vocabulary, activity 3

After the daughter had gone sightseeing, her mother complained to the hotel doctor that she felt ill. She had contracted the plague. The hotel officials were instructed to keep the news quiet in case everyone left the city and the Great Exhibition ended in disaster. The mother's room was quickly cleaned and another guest moved in. No one knows what happened to the mother, but it is assumed she remained in France until she died.

It turned out that the journalist had been so busy looking at the sights he had not noticed that a manhole cover had been removed from the pavement and had simply fallen down it while his friend carried on talking, completely unaware of what had happened.

18 *Lesson 40*
Reading and listening, activity 6

Part 3

Later that night Ray ran out of cigarettes, and when he left the apartment and headed for the corner drugstore, he saw him waiting. Ray knew that this must be the final meeting.

'Say there!'

The Englishman paused and looked at him then turned and walked away from Ray.

'Wait a minute, you! We're going to settle this once and for all.'

But the Englishman kept on walking. Ray followed him down the narrow street to the railroad tracks. He called out, 'Come back! I want to talk to you!'

But by now the Englishman was almost running. Ray started running too, and followed him down the narrow streets that led to the railroad tracks. The Englishman ran on, faster and faster. Finally Ray paused, out of breath. And ahead the Englishman had paused too. Ray saw that he was beckoning him to follow. He broke into a run again. The Englishman waited only a moment and then he too ran, keeping close to the edge of the railroad wall, where only a few inches separated him from a twenty-foot drop to the tracks below.

In the distance, Ray heard the low whistle of the Stamford Express. Ahead, the Englishman was out of sight around the corner for a moment, but Ray rounded the wall and saw, too late, that the Englishman was waiting for him there. The man's big hands came at him and all at once Ray was pushed and falling sideways, over the edge of the railroad wall. And as he hit the tracks, he saw that the Stamford Express was almost upon him, filling all space with its terrible sound...

Grammar review

CONTENTS

Present simple

Use

You use the present simple:

- to talk about something that is regular, such as routines, customs and habits. (See Lesson 2)
  They usually spend Christmas in New York.

- to talk about a general truth, such as a fact. (See Lesson 2)
  Half a million people pass through the station daily.

- to talk about something that stays the same for a long time, such as a state. (See Lesson 2)
  He lives in Queens.

  You can also use the present simple to talk about a personal characteristic.
  She plays the piano.

Present continuous

Use

You use the present continuous:

- to talk about an action which is happening at the moment. (See Lesson 2)
  She's having coffee.

- to talk about an action or state which is temporary. (See Lesson 2)
  He's working at the Met.

- to talk about a definite arrangement in the future. (See Lesson 2)
  They're spending Christmas in Chicago.

- to talk about arrangements with verbs like *arrive come go leave meet see visit* (See Lesson 2)
  We're coming to Chicago by train.

You don't usually use these verbs in the continuous form.
believe feel hate hear know like love smell sound taste understand want

Past simple

Use

You use the past simple:

- to talk about a past action or event that is finished. (See Lesson 6)
  Josephine Wilson trained as an actress.

Irregular verbs

There are many verbs which have an irregular past simple. For a list of the irregular verbs which appear in **Reward Intermediate**, see page 113.

Future simple (*will*)

Use

You use the future simple:

- to make a prediction about the future. (See Lesson 17)
  He'll be here at nine tomorrow.

- to talk about things you are not sure will happen with *perhaps* and *It's possible/probable that...* (See Lesson 17)
  It's possible that I'll be late tomorrow.

- to talk about decisions you make at the moment of speaking. (See Lesson 17)
  I'll give you the money right now.

Remember that you can also use the future simple:

- to offer to do something.
  OK, I'll buy some food.

Present perfect simple

Form

You form the present perfect with *has/have* + past participle. You use the contracted form in spoken and informal written English.

Past participles

All regular and some irregular verbs have past participles which are the same as their past simple form.
Regular: *move – moved, finish – finished, visit – visited*
Irregular: *leave – left, find – found, buy – bought*

Some irregular verbs have past participles which are not the same as the past simple form.
go – went – gone   be – was/were – been
drink – drank – drunk   ring – rang – rung

For a list of the past participles of the irregular verbs which appear in **Reward Intermediate**, see page 113.

Been and *gone*

He's been to America = He's been there and he's back here now.
He's gone to America = He's still there.

Use

You use the present perfect:

- to talk about an action which happened at some time in the past. We are not interested in *when* the action took place but in the experience. We often use *ever* and *never*.
  Have you ever read an English newspaper?

  Remember that if you ask for and give more information about these experiences, actions or states, such as *when*, *how*, *why* and *how long*, you use the past simple.
  When did you read an English newspaper? Last week.

- to talk about a past action which has a result in the present.
  She's done the cooking.

- when the action is finished, to say what has been completed in a period of time often in reply to *how much/many*. (See Lesson 13)
  How many films has James Bond appeared in?

You can use:

- *already* with the present perfect to suggest *by now* or *sooner than expected*. It's often used for emphasis and goes at the end of the clause. (See Lesson 11)
  I'll bring in the rubbish.   I've done that already.

  You put *already* between the auxiliary and the past participle. You don't often use *already* in questions and negatives.

- *yet* with the present perfect in questions and negatives. You use it to talk about an action which is expected. (See Lesson 11)
  I haven't bought the wine yet.
  Have you picked up the children yet?

  You usually put *yet* at the end of the sentence.

- *still* to emphasise an action which is continuing.
  The children are still at school.

  You usually put *still* before the main verb, but after *be* or an auxiliary verb. In negatives it goes before the auxiliary. (See Lesson 11)

Present perfect continuous tense

Form

You form the present perfect continuous tense with *has/have been* + *-ing*. You usually use the contracted form in spoken and informal written English.

Use

You use the present perfect continuous:

- to talk about actions and events which began in the past, continue up to the present and may or may not continue into the future. You use *for* to talk about how long something has been happening.
  I've been learning English for three years.

  You use *since* to say when the action or event began.
  I've been learning English since 1992.

- to talk about actions and events which have been in progress up to the recent past, that show the present results of the past activity. (See Lesson 12)
  It's been raining.

Past continuous

Form

You form the past continuous with *was/were* + present participle. You use the contracted form in spoken and informal written English.

Use

You use the past continuous:

- to talk about something that was in progress at a specific time in the past. (See Lesson 6)
  He was working in a local greengrocer's.

- to talk about something that was in progress at the time something else happened. You join the parts of the sentences with *when* and *while*. The verb in the *when* clause is in the past simple. (See Lesson 6)
  He was staying with Cathy Kelly when he first met Josephine.

  The verb in the *while* clause is usually in the past continuous.
  The phone rang while I was cooking.

- to talk about an activity that was in progress when interrupted by something else.
  I was watching television when there was a knock at the door.

  Remember that you use *when* + past simple to describe two things which happened one after the other.
  She was beautiful and kind to me when we finally spoke.

Past perfect simple

Form

You form the past perfect with *had* + past participle. You use the contracted form in spoken and informal written English.

Use

You use the past perfect simple:

- to talk about an action in the past which happened before another action in the past. The second action is often in the past simple. (See Lesson 35)
  Fame had exhausted Amelia and she decided to retire.

- in reported speech or thoughts.
  'We saw Amelia in Saipan,' people said.
  People said they had seen Amelia in Saipan.

- with *when, after, because* and *until* for the first of two actions.
  Two months after they had got married Amelia flew across the USA coast-to-coast.

  You can use two past simple tenses if you think the sequence of actions is clear.
  She left home in 1920 and went to California.

Past perfect continuous

Form

You form the past perfect continuous with *had been* + present participle. You use the contracted form in spoken and informal written English.

Use

You use the past perfect continuous:

- when you want to focus on an earlier past action which was in progress up to or near a time in the past, rather than a completed event. You often use it with *for* and *since*. (See Lesson 35).
  When she arrived a crowd had been waiting for several hours to greet her.

Questions

Asking questions

You form questions in two ways:

- without a question word and with an auxiliary verb. The word order is auxiliary + subject + verb. (See Lesson 1)
  Are you married?   Was he born in London?
  Do you have any brothers?   Did you get up late this morning?

- with a question word, eg *who, what, where, how, why, when* and an auxiliary verb. The word order is question word + auxiliary + subject + verb. (See Lesson 1)
  What nationality are you?   Where were you born?
  How do you say 'salut'?   What do these words mean?

You can put a noun after *what* and *which*.
What time is it?   Which road will you take?

You often say *what* to give the idea that there is more choice.
What books have you read lately?

You can put an adjective or an adverb after *how*.
How much is it?   How long does it take by car?   How fast can you drive?

You can use *who, what* or *which* as pronouns to ask about the subject of the sentence. You don't use *do* or *did*.
What's your first name?   Who organised the first package trip?

You can use *who, what* or *which* and other question words to ask about the object of the sentence. You use *do* or *did*.
Who did he take on the first package trip?

You can form more indirect, polite questions with one of the following question phrases.
Would you mind explaining the general sense, please?
Could you tell me how you say 'salut'?
I wonder if you could tell me what these words mean?

In the last two question phrases, the word order is question phrase + subject + verb.

Question tags

Question tags turn a statement into a question.

If the statement is affirmative, you use a negative tag. If the statement is negative you use an affirmative tag. (See Lesson 5)
You like jazz, don't you?
You don't think opera is boring, do you?

To ask a real question, the intonation rises on the tag. To show you expect agreement, the intonation falls on the tag. You often use question tags to show friendliness or to make conversation.

Imperatives

You use the imperative for giving instructions. It has exactly the same form as the infinitive (without *to*) and does not usually have a subject. The imperative can be used with a subject to make it clearer who is being spoken to. (See Lesson 26)
Mix the potatoes and onions and cook for two hours.
You mix the potatoes and onions and you cook for two hours.

Verb patterns

There are several possible patterns after certain verbs which involve *-ing* form verbs and infinitive constructions with or without *to*.

-ing form verbs

You can put an *-ing* form verb after certain verbs to describe an activity you either like/don't like doing. (See Lesson 4)
I love walking.
She likes swimming.
They hate lying on a beach.

Remember that *like* and *love* + *-ing* means *enjoy doing something*.
I like going shopping. = I enjoy it.

Use

You use an *-ing* form verb:

- to ask people to do things after *would you mind*. (See Lesson 1)
  Would you mind lending it to me?

- to describe the purpose of something after *to be for*.
  A cassette player is for playing cassettes.

To + infinitive

You can put *to* + infinitive after many verbs. Here are some:
agree decide go have hope learn leave like need offer start try want

Like and *love to* + infinitive suggests that you choose to do something because it's a good idea. You may or may not enjoy it as well.

I like to go shopping on Mondays. = Mondays is the best time for me to go shopping. (See Lesson 4)

Remember that you can use *would like/love to* + infinitive to talk about ambitions, hopes or preferences.

Use

You use *to* + infinitive:

- with *(be) going to* + infinitive to talk about things which are arranged or sure to happen, and for decisions you made before speaking. (See Lesson 17)
  I'm going to arrive on time.

Infinitive of purpose

You use *to* + infinitive to say how or why you do something. (See Lesson 33)
To make your windows really shine, (you can) clean them with wet newspaper.

You can also put the clause of purpose at the end of the sentence.
(You can) clean your windows with wet newspaper to make them really shine.

Need + *-ing* and passive infinitive

You can use *need* + *-ing* form verb or a passive infinitive to say what it is necessary to do. (See Lesson 29)
The switch needs mending. *The switch needs to be mended.*

Causative constructions with *have* and *get*

You can use *have* and *get someone to do something* to mean *ask* or *order someone to do something*. (See Lesson 30)
I'll have an electrician repair it immediately.
We'll get the plumber to do it.

Make and *let*

You can use *make* + noun/pronoun + infinitive to express an obligation. (See Lesson 30)
Make the players play with heavier rackets.
Why don't they make them fight in a round ring?

You can use *let* + noun/pronoun + infinitive to express permission.
I would let the referee look at a video recording.

You can use *not let* to express prohibition.
I wouldn't let them have a second serve.

Infinitive constructions after adjectives

You can use an infinitive (with *to*) after these adjectives. (See Lesson 30)

boring cheap dangerous difficult easy essential expensive good hard important interesting (un)likely (un)necessary (im)possible (un)pleasant (un)usual
It's essential to slow the game down.

You can use *for* + noun between the adjective and the infinitive.
It's very easy for the players to score.

To be due to, to be likely to, to be expected to

You can use *to be due to* + infinitive for something that is arranged. (See Lesson 32)
We are due to receive the parts in a week's time.

You can use *to be likely to, to be expected to* for something that may happen but is not certain.
This is likely to be in a few days.
I am expected to leave hospital in three weeks' time.

By + *-ing*

You can use *by* + *-ing* to say how you do something. The *by* clause can go at the beginning or the end of the sentence. (See Lesson 33)
By adding a cube of plain chocolate to coffee just before using, you can restore the aroma.
You can restore the aroma to coffee by adding a cube of plain chocolate just before using.

Verbs with two objects

Many verbs can have two objects, a direct object and an indirect object. You usually put the indirect object, which often refers to a person, before the direct object. (See Lesson 10)
I gave each guest a box.   He bought me a drink.

You can also put the indirect object after the direct object with a preposition, usually *to* or *for*.
I gave a box to each guest.   He bought a drink for me.

Here are other some verbs which can have two objects.

*cost   lend   make   order   pass   pay   promise   sell
send   teach*

Modal verbs

The following verbs are modal verbs.
can   could   may   might   must   should   will   would

Form
Modal verbs:

- have the same form for all persons.
  I must leave.   He must stay quiet.

- don't take the auxiliary *do* in questions and negatives.
  Can you drive?   You mustn't lean out of the window.

- take an infinitive without *to*.
  I can type.   You should see a doctor.

Use
You use *can*:

- to ask people to do things (See Lesson 1)
  Can you repeat that, please?
  You can also use *could. Can* is a little less formal than *could.*

- to talk about permission (See Lesson 19)
  You can open the window.

- to talk about general ability in the present, something you are able to do on most occasions. (See Lesson 20)
  I can swim very well.

You use *can't*:

- to draw conclusions about things of which you are certain. (See Lesson 18)
  He never sees anyone. He can't like people.

- to talk about prohibition. (See Lesson 19)
  They can't eat the meal in the main restaurant.
  You can also use *mustn't.*
  They mustn't eat the meal in the main restaurant.

- to talk about the absence of a general ability in the present. (See Lesson 20)
  I can't always tell people what I think.

You use *can't have*:

- to talk about something that probably or certainly happened in the past. (See Lesson 38)
  It can't have been a trick.

You can use a verb in its continuous form with past modals.
The voices can't have been speaking in Spanish.

You use *could*:

- to ask polite questions. (See Lesson 1)
  Could I use your pen, please?

- to ask people to do things (See Lesson 1)
  Could you repeat that, please?

- to draw possible conclusions. (See Lesson 18)
  She could be Italian.

- to express general ability in the past. (See Lesson 20)
  When I was young, I could usually get what I wanted.

- to ask for permission.
  Could I leave now?

  Remember that you use *be able to* to talk about ability in the other tenses.
  She'll be able to relax with them.

  You use *was/were able to* to talk about something that was possible on a specific occasion in the past. Here it means *managed to do something.*
  The weather was good so we were able to go out.

  But you can use *couldn't* for general ability and specific occasions.
  He was so busy that he wasn't able to/couldn't come.

You use *could have*:

- to talk about something that possibly happened in the past. (See Lesson 38)
  He could have had an accident.

You use *may*:

- to make predictions about possible future events. (See Lesson 17)
  I may be late tomorrow.

- to draw possible conclusions. (See Lesson 18)
  I may not get there in time.

You can also use *may*:

- to ask for permission.
  May I call you Esther?
  May has almost the same meaning as *might.*

You use *may have*:

- to talk about something that possibly happened in the past.
  He may have got lost. He may have had an accident.

You use *might*:

- to talk about possible future events. (See Lesson 17)
  I might be late tomorrow.

- to draw possible conclusions. (See Lesson 18)
  I might not get there in time.

You use *might have*:

- to talk about something that possibly happened in the past. (See Lesson 38)
  He might have got lost.
  You don't usually use *might* in questions.

You use *must*:

- to give strong advice. (See Lesson 16)
  You must always arrive as conspicuously as possible.

- to draw conclusions about things of which you are certain. (See Lesson 18)
  They have been friends for many years. They must know each other very well.

You can also use *must*:

- to talk about something you are obliged to do. The obligation usually comes from the speaker and it can express a moral obligation, necessity or a strong suggestion.
  It's late. I must go now.   You really must stop smoking.

  You often use it to talk about safety instructions.
  You must fasten your seatbelt.

  Have to has almost the same meaning as *must* but the obligation comes from a third person. You often use it to talk about rules.
  The government says you have to do military service.
  Don't have to suggests absence of obligation.
  You don't have to go to school at the weekend.

You use *must have*:

- to talk about something that probably or certainly happened in the past.
  The journalist must have disappeared for several hours.

You use *mustn't*:

- to give strong advice. (See Lesson 16)
  You mustn't stand on the pavement and wave him goodbye.

You can also use *mustn't*:

- to talk about something you're not allowed to do.
  You mustn't smoke here.

  You can also use *can't*.
  You can't smoke here.

  Remember that you use *have to* in questions and not *must*.

You use *should*:

- to give less strong advice. It can also express a mild obligation or the opinion of the speaker. (See Lesson 16)
  You should put back on your sunglasses.

  Ought(n't) to has the same meaning as *should(n't)*.
  You ought to put back on your sunglasses.

You use *should have* and *shouldn't have*:

- to describe actions in the past which were wrong, or which you now regret. (See Lesson 37)
  She should have gone to Hong Kong in the autumn.

- to criticise someone.
  She shouldn't have worn flip-flops in the restaurant.

  Ought to have and *oughtn't to have* have a similar meaning.

For the uses of *will* see Future simple (*will*)

You use *would*:

- to ask people to do things with *mind* + *-ing* form verb. (See Lesson 1)
  Would you mind explaining the general sense, please?

You also use *would*:

- to ask for permission with *mind* + *if* + past tense although the action is not in the past.
  Would you mind if I left early?

  Remember that you can also use *would* to talk about the consequences of an imaginary situation. (See Lesson 36)
  I'd go to a small town in France.

Zero conditional

Form
You form the zero conditional with *if* + present simple + present simple or imperative.
If I take a spare key, I don't have to disturb the neighbours.

Use
You use a zero conditional:

- to talk about something which is generally true or which usually happens. (See Lesson 31)
  If there's a long wait, I like to have something to read.

- to describe a common problem, followed by an imperative or *you can...* to give advice on what to do about it. (See Lesson 33)
  If lettuce becomes limp, place it in a saucepan with a lump of coal.
  If an oven dish gets burnt, you can put it face down on the grass all night.

  You separate the two clauses with a comma. The *if* clause can go at the beginning or end of the sentence.

First conditional

Form
You form the first conditional with *if* + present simple, *will* + infinitive.
If you speak the language, the locals will be very hospitable.

Use
You use the first conditional to talk about a likely situation and to describe its result. You talk about the likely situation with *if* + present simple. You describe the result with *will* or *won't*. (See Lesson 34)
If you are late, we won't wait for you.

You separate the two clauses with a comma. You often use the contracted form in speech and informal writing. The *if* clause can go at the beginning or end of the sentence.

Second conditional

Form

You form the second conditional with *if* + past simple, *would* + infinitive.
If they got married, it wouldn't last long.

Remember that you can also use the past continuous in the *if* clause.
If it was raining, I'd take an umbrella.

Use

You use the second conditional:

- to talk about an imaginary or unlikely situation and to describe its result. You talk about the imaginary or unlikely situation with *if* + past simple. You describe the result with *would* or *wouldn't*. (See Lesson 36).
  If I won a lot of money, I'd travel around the world.

- to give advice. (See Lesson 36)
  If I were you, I wouldn't let it spoil the relationship.

  You separate the two clauses with a comma. You often use the contracted form in speech and informal writing. The *if* clause can go at the beginning or end of the sentence. It is still common to see *were* and not *was* in the *if* clause.
  If I were you, I'd go home.

Third conditional

Form

You form the third conditional with *if* + past perfect, *would have* + past participle
If I'd seen him, I would have ignored him.

Use

You use the third conditional to talk about an imaginary or unlikely situation in the past and to describe its result. You talk about the imaginary or unlikely situation with *if* + past perfect. You describe the result with *would have* or *wouldn't have*. (See Lesson 40)
If I'd seen him, I would have ignored him.
If he'd spoken to me, I wouldn't have said anything.

You can also use *may have, might have* and *could have* if the result is not certain.

You separate the two clauses with a comma. You often use the contracted form in speech and informal writing. The *if* clause can go at the beginning or end of the sentence.

In case

You use *in case* to give the reason why you do something as a precaution against something that might happen. You use the present simple after *in case* but not a future tense or *going to*.
I always have a bar of chocolate in case I get hungry. – There is a chance that I'll get hungry, so I take some chocolate.

You can also use a past tense after *in case*.
She took a sweater in case it got cold.

If you use *if* instead of *in case* it means you will only do something when it's more certain that something else will happen.
I always take a book if I have to wait for a bus. = On the occasions that I know I will have to wait, I take a book.
I always have a bar of chocolate if I get hungry. = Whenever I get hungry, I always have some chocolate. (See Lesson 31)

I wish and *If only*

You can express wishes about the present to say how you would like things to be different from the way they are with *I wish* + past simple.
I wish I was still a student. = I'm not a student now.

You can express regrets about the past with *I wish* + past perfect. (See Lesson 39)
I wish I had worked harder. = I didn't work hard.

You can usually use *If only* instead of *I wish*.
If only I was still a student. If only I had worked harder.

The passive

Form

You form the passive with the different tenses of *be* + past participle.

Present simple passive: *Spanish is spoken here.*
Past simple passive: *The man was sent to prison.*
Present continuous passive: *The office is being decorated.*
Past continuous passive: *She was being watched by the police.*
Present perfect passive: *The teachers have been paid today.*
Future passive: *It is an industry which will be encouraged.*
Modal passive: *The car might be repaired today.*

Use

You can use the passive:

- when you want to focus on when or where something is/was done, rather than who does/did it. The object of an active sentence becomes the subject of a passive sentence. (See Lesson 28)
  The valleys of Wales are linked with the coal industry.

- to introduce general opinions in an impersonal style. (See Lesson 28)
  It is thought that the most important single influence ...

- to describe processes. (See Lesson 28)
  Apples are brought from the orchards and processed.

  Remember that if you are more interested in the object but you know who or what does something, you use *by*.
  Minton and Spode china is collected by people all over the world.

Used to and would

You use *used to* + infinitive:

● to talk about habits, routines and states in the past. You often use it in narratives. (See Lesson 7)
My uncle used to smoke 40 cigarettes a day.   He didn't use to smoke cigars.

You often use *used to* to contrast past routine with present state.
I used to live in France. I don't live there now. I live in Britain.

You use *would* + infinitive:

● to talk about habits and routines in the past.
The match would always take place ...
You can't use *would* to talk about states.

Describing a sequence of events

Before and after

You use *before* or *after* to link two actions. You use *before/after* + subject + verb when the subject is the same or different in the two actions. (See Lesson 3)
Before they help her dress, they paint her hands and feet.
After he has made sure that the signs are good, the two families ask the gods for their help.

You can also use *before/after* + -ing when the subject is the same in the two actions.
Before helping her dress, they paint her hands and feet.
After arriving at the wedding reception, they sit on a small stage.

You can put the following phrases in front of *before* and *after*.
a week, a day, a year, a few days, shortly, just,
less than a week, for weeks

During and for

You use *during* to say when something happens. (See Lesson 3)
During the wedding, they open the wedding presents.

You use *for* to say how long something takes.
The celebration lasts for several hours.

When, as soon as, as, while, just as, until

You use *when* + past simple for actions which happen one after the other. The second verb can be in either the past simple or the past continuous. (See Lesson 8)
When I saw a group of new arrivals, I wanted to join them.
As soon as I arrived in Moscow, I was going to call my friend.

You can use *when, as* and *while* + past continuous for longer actions. The second verb is often in the past simple.
As we were queuing, a group of soldiers surrounded us.

You can use *just as* + past continuous for shorter actions that happen at the same time. The second verb is often in the past simple.
Just as the plane was landing, he gave me a warning.
Just as I arrived home the phone rang.

You can often use *when* instead of *while/as.*
When we got off the plane, I was feeling a bit nervous.

You use *until* to mean *up to the time when.*
Everything was fine until I came out of customs.

As soon as, when and after for future events

You use the present simple or present perfect to express the future in time clauses after *as soon as, when* and *after.* (See Lesson 32)
Please reply as soon as you receive this letter.
We will be pleased to discuss it after you have recovered.

As soon as means immediately. *When* and *after* are less definite.

Non-defining relative clauses: who, which, where

● You use a non-defining relative clause with *who ,which* or *where* when you add extra information to a sentence. (See Lesson 9)

You use *who* for people.
Van Houten, who was Dutch, was the first person to extract chocolate from cocoa.

You use *which* for things.
Chocolate, which people enjoy all over the world, first came from Central America.

You use *where* for places. Remember that *where* is an object pronoun and must be followed by a subject.
The largest chocolate model was (...) in Barcelona, where they held the Olympic Games...

You put the non-defining relative clause immediately after the person or thing it refers to. The relative pronoun replaces the second noun or pronoun.
Cortes was the first person to bring chocolate to Europe. He was an explorer.

In formal speech, any prepositions go just before the relative pronoun.
The Spanish Court, to which Cortes presented chocolate, drank it with herbs and pepper.

In informal speech, they can go at the end of the clause.
The Spanish Court, which Cortes presented chocolate to, drank it with herbs and pepper.

You separate the relative clause from the main clause with commas, or a comma and a full stop if it is at the end of the sentence.

Defining relative clauses

● You can define people, things and places with a relative clause beginning with *who, that, which, where* or *whose*. The information in the defining relative clause is important for the sense of the sentence. (See Lesson 27)

You use *who* or *that* to define people:

– as a subject pronoun
He's the man who/that does the cooking at the hotel. = He does the cooking.
In this sentence, *who* refers to the subject = *he*.

– as an object pronoun
The nicest person who/that we've met is Spiros. = We met him.
In this sentence, *who* refers to the object = *him*.

You can leave out *who/that* when referring to the object of the relative clause.
The nicest person we've met is Spiros.

You use *who* or *that* to define things:

as a subject pronoun
There's a swimming pool which/that belongs to the house.

– as an object pronoun
It's the nicest place which/that we've ever seen.

You can leave out *which/that* when it is the object of the relative clause.
It's the nicest place we've ever seen.

You use *where* to define places:
This is the village where we're staying.

You use *whose* to replace *his/her* and *their* in relative clauses.
The people whose house we're staying in are interior designers.

You can't leave out *where* and *whose* in relative clauses. Unlike non-defining relative clauses, you don't separate the relative clause from the main clause with a comma.

Making comparisons

You can make comparisons in the following ways:

– comparative adjective + *than*	*Prague is cheaper than London.*
– *more/less* + comparative adjective + *than*	*Rome is more beautiful than London. London is less dangerous than New York.*
– *more* + countable/uncountable noun + *than*	*Tokyo has more inhabitants than Madrid. Berlin has more rain than Rome.*
– *fewer* + countable noun + *than*	*Madrid has fewer tourists than Paris.*
– *less* + uncountable noun + *than*	*Paris has less industry than Milan.*
– *as many* + countable noun + *as*	*Milan has as many people as Barcelona.*
– *as much* + uncountable noun + *as*	*Barcelona has as much rain as Rome.*
– *as* + adjective + *as*	*Rome is as beautiful as Prague.*

Adjectives

Adjectives ending in *-ed* and *-ing*

Many adjectives formed from past participles describe a feeling or a state. (See Lesson 5)
I'm bored. = I feel there is nothing that interests me.
Many adjectives formed from present participles describes the person, thing or topic which produces the feeling.
I'm boring. = I'm a very uninteresting person.
Football is very exciting. = Football excites me.

Order of adjectives

When there is more than one adjective before a noun, you usually put personal opinions before objective facts.
(See Lesson 15)
An efficient, electric alarm clock. My wonderful, brand new mountain bike.

With a series of adjectives, the order is, opinion, size, age, shape, colour, origin, material, purpose, noun and other things.

When there is a list of adjectives, you usually put a comma after each one except the one before the noun.

Possessive adjectives

Form

my - mine   your - yours   his - his   her - hers   its - its
our - ours   their - theirs

Use

You use a possessive pronoun when the noun is understood.
(See Lesson 15)
It was my grandmother's ring, but it's mine now.

Pronouns

Subject	Object	Possessive (See Lesson 15)	Reflexive (See Lesson 29)
I	me	mine	myself
you	you	yours	yourself yourselves
he	him	his	himself
she	her	hers	herself
it	it	its	itself
we	us	ours	ourselves
they	them	theirs	themselves

Adverbs

Formation of adverbs

You form adverbs by adding *-ly* to most adjectives.
(See Lesson 21)
amazing   amazingly   beautiful   beautifully

With adjectives ending in *-y* you drop the *-y* and add *-ily*.
extraordinary   extraordinarily   clumsy   clumsily

With adjectives ending in *-le* you drop the *-e* and add *-y*.
remarkable remarkably terrible terribly

Position of adverbs and adverbial phrases

An adverb or adverbial phrase of manner describes how something happens. (See Lessons 4 and 22)
He raised his head slowly.

An adverb or adverbial phrase of place describes where something happens.
He was drinking at a waterhole.

An adverb or adverbial phrase of time describes when something happens.
Then he turned his back on us.

Adverbs and adverbial phrases usually go after the direct object. If there is no direct object, they go after the verb.
He disappeared quickly.

If there is more than one adverb or adverbial phrase, the usual order is manner, place, time.
We had set out by jeep after supper. (manner, time)
He (...) disappeared quickly into the jungle. (manner, place)

But some adverbs and adverbial phrases can also go before the verb clause if you want to emphasise them.
Twenty years ago, the tiger was in trouble.

Here are some common adverbial phrases of frequency. They usually go at the end of a clause.

every   *day/week/month/year*
one a   *day/week/month/year*
most   *days/weeks*

There are other adverbial phrases that can be put at the end of a clause.
month, year, other days, two days, three times a year, once in a while, every now and then.

Emphasising

You can also use the following adverbs of degree before an adjective to emphasise something. (See Lesson 21)
absolutely amazingly extremely especially extraordinarily particularly really
It's an especially exciting film.

Reported speech

Reported statements

You report what people said by using *said (that)* + clause. You often change the tense of the verb in the direct statement 'one tense back' in the reported statement. Time references also change. The tense change is as follows:

present simple ——▶ past simple
present continuous ——▶ past continuous
present perfect ——▶ past perfect
past simple ——▶ past perfect

However, in spoken language, it's common not to change tenses in the reported statement, especially if the statement is still true at the time of reporting. (See Lesson 23)

'I'll always love you,' she said.   *She said she will always love him.*

Reported questions

You report questions by using *asked* + clause. You usually change the tense of the verb in the reported clause into a past tense. (See Lesson 24)

Direct questions	**Reported questions**
With question words (*who, what, how, why* etc)	
'Which newspaper has the largest circulation?' he asked.	*He asked which newspaper had the largest circulation.*
'How many channels are there?' he asked.	*He asked how many channels there were.*
Without question words *if* or *whether* is used.	
'Is satellite television very popular?' he asked.	*He asked if satellite television was popular.*

Remember that a reported question is not a direct question, so the word order is not the word order of a question. The auxiliary *do* and *did* is not used.
Where do you work?   *She asked me where I worked.*

Make sure you use the correct punctuation.

Reporting verbs

You can avoid repeating what someone said by using a reporting verb to describe the general sense. There are several patterns for these verbs. (See Lesson 25)

Pattern	**Verbs**
1 verb + *to* + infinitive	*agree ask decide hope promise refuse*
Rosemary decided to take the girl home.	
2 verb + object + *to* + infinitive	*advise ask encourage persuade remind warn*
Rosemary persuaded the girl to come home with her.	
3 verb + (*that*) clause	*agree decide explain hope promise suggest warn*
The girl replied that she didn't have any money.	
4 verb + object + (*that*) clause	*advise tell warn*
Jim told her that he didn't want to go.	
5 verb + object	*accept refuse*
She accepted the invitation.	
6 verb + 2 objects	*introduce offer*
She offered her a cup of tea.	

Irregular Verbs

Verbs with the same infinitive, past simple and past participle

cost	cost	cost
cut	cut	cut
hit	hit	hit
let	let	lct
put	put	put
read /ri:d/	read /red/	read /red/
set	set	set
shut	shut	shut

Verbs with the same past simple and past participle, but a different infinitive

bring	brought	brought
build	built	built
burn	burnt/burned	burnt/burned
buy	bought	bought
catch	caught	caught
feel	felt	felt
find	found	found
get	got	got
have	had	had
hear	heard	heard
hold	held	held
keep	kept	kept
learn	learnt/learned	learnt/learned
leave	left	left
lend	lent	lent
light	lit/lighted	lit/lighted
lose	lost	lost
make	made	made
mean	meant	meant
meet	met	met
pay	paid	paid
say	said	said
sell	sold	sold
send	sent	sent
sit	sat	sat
sleep	slept	slept
smell	smelt/smelled	smelt/smelled
spell	spelt/spelled	spelt/spelled
spend	spent	spent
stand	stood	stood
teach	taught	taught
understand	understood	understood
win	won	won

Verbs with same infinitive and past participle but a different past simple

become	became	become
come	came	come
run	ran	run

Verbs with a different infinitive, past simple and past participle

be	was/were	been
begin	began	begun
break	broke	broken
choose	chose	chosen
do	did	done
draw	drew	drawn
drink	drank	drunk
drive	drove	driven
eat	ate	eaten
fall	fell	fallen
fly	flew	flown
forget	forgot	forgotten
give	gave	given
go	went	gone
grow	grew	grown
know	knew	known
lie	lay	lain
ring	rang	rung
rise	rose	risen
see	saw	seen
show	showed	shown
sing	sang	sung
speak	spoke	spoken
swim	swam	swum
take	took	taken
throw	threw	thrown
wake	woke	woken
wear	wore	worn
write	wrote	written

Pronunciation guide

/ɑː/	park	/b/	buy
/æ/	hat	/d/	day
/aɪ/	my	/f/	free
/aʊ/	how	/g/	give
/e/	ten	/h/	house
/eɪ/	bay	/j/	you
/eə/	there	/k/	cat
/ɪ/	sit	/l/	look
/iː/	me	/m/	mean
/ɪə/	beer	/n/	nice
/ʊ/	what	/p/	paper
/əʊ/	no	/r/	rain
/ɔː/	more	/s/	sad
/ɔɪ/	toy	/t/	time
/ʊ/	took	/v/	verb
/uː/	soon	/w/	wine
/ʊə/	tour	/z/	zoo
/ɜː/	sir	/ʃ/	shirt
/ʌ/	sun	/ʒ/	leisure
/ə/	better	/ŋ/	sing
		/tʃ/	church
		/θ/	thank
		/ð/	then
		/dʒ/	jacket

Tapescripts

Lesson 1   Listening, activity 1

GAIL   You want to use your friend's pen. What do you say?

STUDENT   Er, give me your pen. Maybe not, um … I'd like your pen. Yeah, can I use your pen, please?

GAIL   Yes. Question 2. You don't know what an English word, for example, *chart* means. What do you say?

STUDENT   What does *chart* mean?

GAIL   Yes.

STUDENT   Can I go now? [laughter]

GAIL   No, you can't. Question 3. There are over twenty words in a lesson which are new to you. What do you say to your teacher?

STUDENT   Could you explain what they mean? Or … um … when does this lesson finish?

GAIL   No.

STUDENT   No, I suppose, which words should I write down?

GAIL   Yes. Yes, excellent. Question 4. Your teacher says something very quickly to you. What do you say?

STUDENT   What! No, no … slow down … I suppose not. Could you repeat that, please?

GAIL   Yes, that was perfect. Question 5, then. You don't know how to say something, for example, *salut* in English. What do you say?

STUDENT   How do you say *salut* in English?

GAIL   Yes, perfect. Question 6. You don't know how to spell an English word, for example, *sentence*. What do you say?

STUDENT   Um … how do you spell *sentence*?

GAIL   Yes, yes.

STUDENT   But you, how … no … no … that's right.

GAIL   That's perfect. Question 7. Your teacher asks you to read a difficult passage. What do you say?

STUDENT   Could you pass me the dictionary? Um … I wonder if you could tell me what these words mean?

GAIL   No.

STUDENT   No, all right, that has to be *c*, then. Would you mind explaining the general sense of the passage?

GAIL   And question 8. Your teacher plays you a listening text which is very difficult. What would you say?

STUDENT   It's too difficult.

GAIL   No.

STUDENT   That leaves, could you write up some important words and then play it again, please?

GAIL   Yes.

Lesson 2   Vocabulary and listening, activity 1

TICKET-SELLER   Yeah, lady.

JOANIE   Two round trips to Chicago, please.

TICKET-SELLER   There you go. That'll be $192 each. $384 in all.

JOANIE   Would you tell me what track the three o'clock train leaves from?

TICKET-SELLER   Er … let me see, now, that'll be track five.

Lesson 2   Vocabulary and listening, activity 3

Interview 1

Q   Do you often come to Grand Central Station?

FRAN   Yes, every day. I come in from Poughkeepsie.

Q   Where's that?

FRAN   Upstate. It takes about an hour on the train.

Q   Are you waiting for a train now?

FRAN   No, I'm not. I'm going to work. I'm having a cup of coffee and something to eat. Then I'm off to work.

Q   Do you often stop off for coffee?

FRAN   Most days, sure.

Q   May I ask, er … what do you do?

FRAN   I work for an advertising agency. I'm an account executive.

Q   Where do you work?

FRAN   Near Central Park, a few blocks along Madison Avenue.

Q   And how do get there from here? Do you take the subway?

FRAN   No, I usually walk. It's not far from here.

Interview 2

Q   Excuse me, are you in a hurry?

HENRY   Well, we're having a rehearsal in about ten minutes and I'm late. I usually walk but it's raining so I'm taking a cab.

Q   Can I just quickly ask you what you do and where you're going?

HENRY   Well, usually I teach music but at the moment I'm playing cello for the orchestra at the Met.

Q   The Met?

HENRY   The New York Metropolitan Opera.

Q   Oh, and that's where you're going to now?

HENRY   That's right. The rehearsal starts at 2 o'clock.

Q   Is it far?

HENRY   No, just a couple of blocks. About five minutes' walk.

Q   And where do you live?

HENRY   In Queens, in the suburbs. It takes ten minutes by train.

Q   And do you pass through Grand Central every day?

HENRY   Yes, most days.

Interview 3

Q   Excuse me, are you travelling by train today or waiting for someone to arrive?

STEPHEN   We're taking the three o'clock train to Chicago.

Q   And what are you going to do in Chicago?

JOANIE   We're spending Christmas with our grandchildren. They usually spend it with us in New York, but this time we're going to them.

Q   So, a family Christmas in Chicago.

STEPHEN   Well, we're spending Christmas in Chicago and then we're taking the train to Toronto, where we're staying a few days with friends.

Q   And what time do you get to Chicago?

JOANIE   At nine in the morning.

Q   And when are you leaving for home at the end of the vacation?

STEPHEN   When our friends get bored of us.

JOANIE   Ignore him, he's just kidding. On New Year's Eve, 31st of December.

Q   And what do you do?

JOANIE   We're senior citizens.

STEPHEN   I'm a retired stockbroker. I worked in Wall Street for forty years.

Q   And do you live in New York?

STEPHEN   No, sir. We live on Staten Island, and we're proud of it.

Lesson 4   Listening, activity 1

SPEAKER 1   I spend most of my time doing odd jobs around the house. I really like painting and decorating, although I can do most things. If I run out of things to do at home, I usually offer to help the neighbours. I hate having nothing to do.

SPEAKER 2   Well, I go to the match on Saturday afternoon, and then go out with my mates for a drink in the evening. If it's a home game, I can usually get back to watch Match of the Day on television. Then on Sunday we all meet up for a game of football. I belong to the Southfield Superstars, which is part of the Southern Sunday League. We play every Sunday from September to May.

SPEAKER 3   We always have people for dinner on Saturday nights, or we get invitations to have dinner with friends. Sometimes we go out to restaurants. We like Chinese food, and there are also plenty of good Indian restaurants.

SPEAKER 4   I go shopping. I adore it and I go shopping for clothes at least twice a month. And if I haven't got any money, I go window shopping and decide what I'm going to buy when I'm rich.

SPEAKER 5   In the summer, I watch the cricket. I adore it, it's a very relaxing game. I like watching local matches on the village green, but I also like to go to the Test matches if I can. And if I can't, I listen to the radio commentary on the BBC.

SPEAKER 6   I go round the clubs, meeting friends, dancing, listening to music, that sort of thing. I try to go out every night if I can, because, I can't stand staying at home. I usually stay out until about, two or three, unless I'm going to work. Then I try to be in bed by midnight.

SPEAKER 7   I go down to the river at least once a week, usually on Sunday because that's when I want to get away from the children at home. I don't mind spending Saturday with the children, but I get bored if it's the whole weekend. My wife doesn't mind. She says she prefers me out of the house. But she's pleased when I catch enough fish for supper.

SPEAKER 8 I spend a lot of time there now that I'm retired. There's always something to do, cutting the grass, weeding the flower beds, planting bulbs. I like the spring best of all – not much to do and lots to look at. But I detest collecting the leaves in autumn.

SPEAKER 9 In the evenings we watch television, mostly. We don't go out much because our children are still quite young. We've got satellite TV so there's plenty of programmes to watch. And it gives you something to talk about when you're at work the next day.

Progress check 1–5   Sounds, activity 3

Conversation 1
MAN 1 Did you have a good time?
MAN 2 Yes, it was great.
MAN 1 Who won then?
MAN 2 We did. They played really well.
MAN 1 Who scored?
MAN 2 Matthews in the first half and then Jones ten minutes from time. One of the best goals I've ever seen - certainly the best this season.

Conversation 2
MAN Excuse me, but I was waiting to park my car there.
WOMAN Well, that's too bad. I got here first.
MAN I'm sorry, but you can't do that!
WOMAN Well, who says I can't. Are you going to make something of it?
MAN I think it's really very rude of you just to push in like that. You saw me waiting for that space.
WOMAN Too bad.

Conversation 3
MAN Excuse me, but, er … could I go before you please? I'm in a bit of a hurry.
WOMAN Well, actually so am I.
MAN I'm sure you won't mind. I've only got a few things and my car is badly parked outside and … and I've got to pick the children up from school.
WOMAN Well, all in good time. I've got to get home as well.
MAN Oh, please, I am really in a hurry …
WOMAN Yeah …

Lesson 6   Listening and writing, activity 2

Q So how long have you known each other?
JOE For ever!
SARA It's seems like ages.
JOE Well, it's not that long. I've only been in England for about six months, and we met just before last Christmas.
Q And where did you meet?
SARA It was at an end-of-term party at the college we both go to. Everyone was happy and relaxed and looking forward to Christmas.
JOE Hey, wait a minute! *You* were happy and relaxed but I was feeling pretty miserable, because I'd said goodbye to all my friends who had gone back to the States to be with their families for Christmas, and, I was planning to spend it on my own here.
Q So what were you doing at the party?
JOE Well, there were a lot of people there, and I didn't really know anybody, so I was standing around with no one to talk to, and thinking about going back to my flat …
SARA And I was talking to some friends, chatting about our holiday plans, when I saw Joe looking, well, erm …
JOE Moody and magnificent?
SARA Well, not really, but I did say to my friend, 'Who is that?' and she said, 'Come on over I'll introduce you to him.'
JOE And we spent the rest of the evening chatting.
Q And why did you become friends?
SARA Well, Joe has a great sense of humour, which I like a lot in men. And we discovered that we had similar interests and erm … taste in music, that kind of thing.
JOE So I asked if I could see her again, but she said she was leaving next day for the Christmas vacation.
SARA Then, the next day I rang him to say that I wasn't leaving until the end of the week, and that it would be nice if we could meet.
JOE And it was the best Christmas present I could have had.

Lesson 8   Vocabulary and listening, activity 3

Part 1
FRIEND So, how was Russia? How did you get on?
CLAUDIA Well, it was quite an experience. I started my trip in Moscow. The trouble was I forgot to ring my friend Sergei before I left London. Then, just as the plane was landing at Moscow airport, this German businessman sitting next to me called Stefan, gave me a warning, "You shouldn't arrive in Moscow alone." This made me a bit worried, but I was going to call Sergei as soon as I arrived, so I hoped I would be OK. But when we got off the plane I was feeling a bit nervous. And then, as we were queuing at passport control, a group of soldiers surrounded us, and I stayed very close to Stefan. Everything was fine until I came out of customs. As soon as I walked out into the hall this crowd of people surrounded me. And when Stefan disappeared, I felt lonely and rather foolish.
FRIEND I can see why.
CLAUDIA And when I saw a smiling group of new arrivals, you know, package tourists, I wanted to join them. Just as I was starting to panic, someone tapped me on the shoulder. It was Stefan. I was so relieved.
FRIEND Thank heavens for strange men!
CLAUDIA Exactly! But when I called Sergei from Stefan's office, he was out. It was late by now so I had to find a hotel. Stefan told his driver to take me to a hotel near Arbat Street. As soon as I got out of the car, three soldiers stopped me with rifles. When they heard me speaking English, they let me through. And at the reception desk when I explained that I was on my own, they didn't want to give me a room.
FRIEND So what did you do?
CLAUDIA Well, I had to wait at reception until the theatre in St Petersburg, where I was going to work, sent a fax. When the hotel knew what I was doing in Russia, I got a room. Things definitely work differently in Russia.

Lesson 8   Vocabulary and listening, activity 5

Part 2
CLAUDIA Well, in the end I got in touch with Sergei and I spent the next few nights with his family. And while I was staying with them, I even learnt the Russian alphabet and a few words. But it was cold, and a freezing cold wind was blowing all the time.
FRIEND But it all went better for you after that?
CLAUDIA Well, yes, until one morning I decided to take Max, the dog, for a walk. I was really enjoying the walk, when I realised that I was lost. Then I found a playground, which looked like the one outside the flat … but then I found another one, and another one.
FRIEND Oh no! You must have been very frightened!
CLAUDIA Well, I was. I showed the address to various people, but didn't understand their replies. It was freezing cold, minus 12 degrees. I was standing by the side of a road, praying that Sergei would send out a search party, when suddenly a car stopped and the driver said, "Taxi?" I was so relieved. Within a couple of minutes I was drinking tea with Sergei and the driver back home. His name was Nikolai and he was on his way to work. He wasn't a taxi driver at all.
FRIEND Was he a friend of Sergei's?
CLAUDIA No, just someone who was going to work when he saw a foreigner looking lost, I think. They can be very friendly, the Russians.

Lesson 10   Listening, activity 1

WOMAN So what about Lu, from China. What did he do wrong?
MAN Well, Lu obviously was aware that things were different in Britain, um … in China the signal that a meal is over is when the last dish is served. Um, he being sensitive to things being different in Britain, tried to do the right thing but unfortunately asked the question in a rather clumsy way which gave the impression that he wanted to leave very soon.
WOMAN Yes, now what about Douglas from Scotland and his choice of flowers for the hostess?
MAN Well, traditionally you should always give an odd, rather than an even number of flowers, so in the first place twelve was the wrong number, and also roses suggest a relationship which is more than just friendly, it suggests love and intimacy, so roses were perhaps the wrong choice. So wrong number and wrong type of flowers really. The wine was probably OK as a gift.
WOMAN Now what about Kenji from Japan, who seemed to expect that his host would pay for his meal?

MAN Well, there's a phrase in English called 'going Dutch', um ... which is fairly common, when you go out for a meal with friends, each person tends to pay for his or her share of the bill. Unless you're being specifically taken out for a meal, in which case one person will, but it is quite common to 'go Dutch' as they say, share the bill.

WOMAN Yes, and Carlos from Spain. Um ... confused on the bus, what would you explain to him?

MAN Well, it's very common to expect younger people in Britain to give up their seat to older people. At least that's what older people think, unfortunately not every British young person knows about this.

WOMAN No, and Hannah from Lebanon, um ... feeling confused about, um ... how far to go maybe in an introductory friendship.

MAN Well, people in the United States tend to be very outgoing, and very friendly ...

WOMAN And very open.

MAN Very open on the surface, um ... they like to make you feel special when they're talking to you, and they try to be as friendly as they possibly can. But, as in all other countries it always takes a long time for people to really make relationships and really make close friends and so the depth of the relationship underneath takes a longer time than the surface friendship.

WOMAN Yes.

Lesson 11   **Listening and vocabulary, activity 2**

IAN Hi, Kate. How are you getting on?

KATE Hey, what are you doing here? I didn't expect you until later.

IAN Well, I've already finished everything I had to do at work, so I thought I'd come back and help you. I know you've had a busy day.

KATE I've had an extremely busy day, and it hasn't finished yet. Did you remember we've got Paul and Hannah for dinner tonight?

IAN Yes, I did. Have you got everything you need?

KATE Well, I've already done the shopping, but I haven't bought the wine yet.

IAN Oh, I'll do that. Have you picked up the children yet?

KATE No, I haven't done that yet. They're still at school.

IAN OK, well, I'll get them when I buy the wine.

KATE No, they've got their sports club, so they'll be there until about six. You're. more important here. Have you collected the car from the garage yet? It should be ready by now.

IAN No ... No, I haven't. I'll ... I'll go to the garage later. Now, what's next?

KATE Well, er ... the kitchen is clean, but you could check the bathroom for me.

IAN OK. Have you thrown away the pile of newspapers in the front room?

KATE No, they're still lying on the table.

IAN I'll take out the rubbish as well.

KATE I've already done that. Now, I haven't started cooking yet, and it's getting late. When you've got the wine, can you lay the table? Here's the table cloth and knives and forks. And mend the television? It still doesn't work. Oh, and wrap up your mother's birthday present. It still needs wrapping paper. And turn on the heating. It's getting quite cold in here.

IAN Oh dear, it's one of those days.

Lesson 12   **Vocabulary and listening, activity 2**

Good evening. Here is the six o'clock news. Hijackers are still holding twenty-three passengers in a plane at Manchester airport. They hijacked the flight from London to Glasgow last Thursday. The hostages have now been sitting in the plane without food or water for three days. The hijackers want the release of some political prisoners, but the government refuses to give in.

Planes are now landing in the Balkans with supplies, after flights were resumed earlier today, following the end of the civil war. The people in the Balkan capital have been waiting for food for the last six months. It is hoped that life in the region will return to normal within a couple of months, although it will be many years before they can repair the structural and emotional damage of the war.

Traffic in west London has been growing steadily since early this morning as fans begin to arrive for today's football match at Wembley stadium where the England team is playing Brazil this afternoon. Police are advising motorists to avoid the area unless absolutely necessary.

A demonstration against unemployment has been taking place in Manchester. Demonstrators have been marching through the city for two hours. It is expected to finish in front of the town hall at five this evening where left-wing Members of Parliament will address the crowd.

The fire on the oil rig in the North Sea is still burning after the explosion last Tuesday. Helicopters took the crew to safety, but firefighters have not been able to get close to the source of the fire. Bad weather is causing problems with gales and heavy seas.

Despite good weather over most of the country, Scotland has been experiencing some of the worst summer weather this century. It has been raining there for the last ten hours, and the forecast suggests that it will continue until tomorrow morning, with a chance of thunder and lightning.

Lesson 14   **Vocabulary and listening, activity 4**

Q I've never actually been to Prague so what's it like I've no idea – is it a big city?

RICHARD Um, I think it's l.2 million - obviously a lot smaller than London.

Q Oh, right, because we've got 7 million here I'd say so, um ... London's a lot bigger?

RICHARD Even more I'd say, now.

Q I mean, how many of those are tourists? How many tourists go there, in fact?

RICHARD Well, obviously now it's a very popular city. There's loads of tourists probably equal to London I'd say.

Q Is there much industry there? Is it an industrialised city?

RICHARD Um ... fortunately not. No, not really. There's a little on the edge of the city but nothing at all really in the centre.

Q So, it sounds like there's a lot less than in London.

RICHARD Oh, yes, absolutely.

Q What's the weather like?

RICHARD Well, it's pretty extreme really. It's very hot in the summer and um ... very cold in the winter.

Q So, you don't have rain, like we have. We've got er ... London has rain in the winter and it's pretty rainy in the summer as well. Is it an expensive city?

RICHARD Well, because of the exchange rate, I think it's quite cheap for British people, I mean ... a beer costs ... what, fifteen pence.

Q Fifteen! Do you know how much they are here?

RICHARD No.

Q One pound, fifty.

RICHARD You're joking!

Q Yes, very expensive, London. Um ... I've no idea what it looks like. What's the architecture like ... is it a nice place?

RICHARD Oh, it's beautiful. Full of beautiful, fantastic buildings. I mean, wonderful baroque and gothic buildings.

Q We have those too, but we ... some parts of London are very nice but it also has some very run down areas which are not so nice.

RICHARD Well, because it's bigger.

Q Yes, I suppose so, and also parts of it are very, very dirty in London. I don't know ... what is ... is...?

RICHARD No, actually, I have to say it's a very clean city. There's hardly any rubbish at all on the streets.

Q Do you feel safe there walking around, and travelling around?

RICHARD Yes, absolutely. I mean, a lot safer than I used to feel in London. Um, I mean ... um ... you know at night, you feel absolutely fine.

Q Well, that's the thing, isn't it?

RICHARD Well, let's hope it doesn't change, that's all.

Q London at night is, well ... it's OK during the ... day, at night. It can get nasty anyway.

RICHARD So I hear.

Q So, what's the kind of atmosphere of the place?

RICHARD It's very romantic.

Q Is it?

RICHARD Yeah, in a word that sums it up. It's wonderful at night, by the river you know ... looking up at the castle on the hill there. It's quite fabulous..

Q Yeah, that must be good. I would never call London really romantic. It's kind of lively ... you know, it has a life of its own.

RICHARD Yeah.

Lesson 17   **Listening and speaking, activity 3**

Rose Rose, part 1

Sefton stepped back from his picture. 'Rest now, please,' he said.
Miss Rose Rose, his model, threw the striped blanket around her, and seated herself on the floor near the big stove. For a few moments Sefton stood motionless, looking, critically at his work. Then he laid down his palette and brushes, and began to roll a cigarette.

'Well,' he said, 'I've got the place hot enough for you today, Miss Rose.'

'You have indeed,' said Miss Rose. 'The trouble with these studios is the draughts.'

She was extremely beautiful. Every pose she took was graceful. She knew her work well and did it well. On one occasion, when sitting for the great painter Merion, she'd kept the same position, without a rest, for three hours. But you could not absolutely depend upon her. Sometimes she kept her appointments, sometimes she arrived an hour or two late, or sometimes she didn't arrive at all. Sefton continued to work hard for half an hour in silence. Then, his eyes felt tired and he said he could not see any longer.

'We'll stop for today,' he said. Miss Rose Rose retired behind the screen.

'Tomorrow, at nine?'

'I could be here by a quarter past.'

'Right,' said Sefton, as he put on his coat.

When Rose Rose came out from behind the screen she was wearing a blue dress and a hat.

'I say, Mr Sefton,' said Rose, 'I know you said you'd pay me at the end of the sittings, but -'

'Oh, you don't want any money. You're known to be rich,' he joked.

'I know what you are thinking, Mr Sefton. You think I don't mean to come tomorrow. That's because of Mr Merion, isn't it? He's always saying things about me. And all because I was once late - with a good reason for it, too.'

'Well, here's your money, Miss Rose. But will you really be here in time tomorrow?'

'You needn't worry about that,' said Miss Rose, eagerly. 'I'll be here by a quarter past nine. I'll be here if I die first! Good afternoon, and thank you, Mr Sefton.'

'Good afternoon, Miss Rose.'

Sefton didn't feel sure of her, but it was impossible for him to refuse to pay her what he really owed. He picked up his hat and went out. He walked the short distance from his studio to his flat, then changed his clothes and took a cab to the club for dinner. He played a game of billiards after dinner, and then went home. His picture was very much in his mind. He wanted to be up early in the morning, so he went straight to bed.

Lesson 17 Listening and speaking, activity 5

Rose Rose, part 2

He was at his studio by half-past eight. His painting of Aphrodite seemed to have a mocking expression. He then realised he didn't have any tobacco. It was still a few minutes before nine. There was time to go up to the tobacconist on the corner. In case Rose Rose arrived while he was away, he left the studio door open. He bought a morning paper. Rose would probably be twenty minutes late at the least. But on his return he found his model was already there.

'Good morning, Miss Rose. You've kept your promise.' He threw his cigarette into the stove, picked up his brush, and started work. For some time he thought of nothing else. There was no relaxing on the part of the model, no sign of fatigue.

'We'll have a rest now, Miss Rose,' he said cheerily. At the same moment he felt human fingers drawn lightly across the back of his neck. He turned round. There was nobody there. He turned back again to his model. Rose Rose had disappeared.

With great care he put down his palette and brushes. He called, 'Where are you Miss Rose?' His voice sounded loud and frightened. He repeated his question, but there was no reply. Nervously, he stepped behind the screen, and suddenly he realised that the model had never been there at all.

He walked over to the back door – he had locked it on his return from the tobacconist. The door was still locked. Slowly, he went back, sat down in the chair, and lit a cigarette. Undoubtedly he had been working very hard lately, but he was not aware of behaving strangely. He did not feel ill. But at this moment he was absolutely terrified. With a great effort he tried to behave normally. He picked up the newspaper and started to read. It was certain he could do no more work for that day, anyhow. But his model, Miss Rose, he really believed she had been there.

He read the newspaper carefully. Then his eye fell on a paragraph headed 'Motor Fatalities'. He read that Miss Rose, an artist's model, had been knocked down by a car in the Fulham Road about seven o'clock on the previous evening; that the owner of the car had stopped and taken her to hospital, and that she had died within a few minutes of arriving there.

Sefton sat motionless for a few moments. And then he rose from his seat and, opening a large pocket-knife, he walked slowly towards the picture. He looked at the face in the painting smiling at him. He felt strangely calm, but the picture …

Lesson 18 Listening

Q	Do you work inside?
MAN	Yes.
Q	You do, so you're not outside. Do you work at home?
MAN	No.
Q	You don't. Do you work at weekends?
MAN	No.
Q	Right. Er … do you need to be … attractive?
MAN	Er … no.
Q	Do you need to be ambitious?
MAN	Hmm, yes
Q	Right. Do you need to be physically fit?
MAN	Yes.
Q	Um … do you need to … work with your hands?
MAN	No.
Q	Do you need to make or build anything?
MAN	No.
Q	Um, do you need to wear a uniform?
MAN	No.
Q	Do you have to talk a lot?
MAN	Yes.
Q	Um … do you have to have special qualifications?
MAN	Yes.
Q	Do you have to give orders or instructions to others?
MAN	Yes.
Q	Could you possibly be a teacher?
MAN	Yes, I am. Well done.

Q	Do you work outside?
WOMAN	No, never.
Q	Right, so you work inside. Um … do you work in an office?
WOMAN	No.
Q	No. Do you work at weekends?
WOMAN	Oh, yes.
Q	Right, um … do you need to be ambitious?
WOMAN	Yes.
Q	Do you need to be attractive?
WOMAN	Um … yes, it helps.
Q	Do you need to be physically fit?
WOMAN	Not really, no.
Q	Right, um … do you wear a uniform?
WOMAN	Um … no … no.
Q	Do you work with your hands?
WOMAN	Yes.
Q	Do you use a machine?
WOMAN	Oh, yes.
Q	Ah, do you have to meet a lot of people?
WOMAN	Yes, lots.
Q	And do you have to have special qualifications?
WOMAN	Yes.
Q	Do you talk a lot?
WOMAN	Oh yes.
Q	Could you possibly be a hairdresser?
WOMAN	Yes. Well done

Q	Do you work at home?
MAN	No, no I don't.
Q	Do you work in an office?
MAN	No.
Q	Do you work outside?
MAN	Yes.
Q	Do you need to be imaginative?
MAN	Um … yes.
Q	Do you need to be physically fit?
MAN	Yes, definitely, yeah.
Q	Right. Do you need to be ambitious?
MAN	Not really, no.
Q	No. Do you need to work with your hands?
MAN	Yes.
Q	And, do you use a tool?
MAN	Yes, yup.
Q	Do you use a machine?

MAN	Yes, yes.
Q	Right. Do you, er … have to have special qualifications?
MAN	Um … no.
Q	Do you meet a lot of people?
MAN	Um … not really, no.
Q	Do you give orders or instructions to others?
MAN	Um … sometimes, but not really, no.
Q	Are you a builder?
MAN	No, no, close, I'm a gardener.

Lesson 20 **Listening and writing, activity 1**

Part I

FRANK	What are you smiling about?
ROSALIND	Well, you're very lucky I'm smiling, I can tell you. I could be crying.
FRANK	Why? What's happened?
ROSALIND	I have just lost my temper.
FRANK	Oh dear.
ROSALIND	Well, I think I ought to do it more often.
FRANK	What happened?
ROSALIND	I was having lunch today with my friend Peter. I drove round to his flat and picked him up, then we drove into the town centre and found a parking space in the multi-storey car park in George Street. We had lunch in a restaurant nearby and then went back to the car. I paid for the parking and got my exit ticket. We couldn't take the stairs because Peter has a bad leg and couldn't walk very far, so we got into the lift. The doors shut, and the lift started moving. Then, suddenly, there was a loud noise and the lift stopped.
FRANK	Oh no.
ROSALIND	You guessed it.
FRANK	Stuck?
ROSALIND	The lift was stuck! And the lights went out. Eight of us stuck between floors four and five. One of the men tried to open the door. He was able to open it a few inches, but then it shut again, automatically. I pressed the alarm bell, but we couldn't hear it. We wondered if it was working.
FRANK	Were you frightened?
ROSALIND	Well, not really. At first, it was rather boring. We were able to talk, of course, but we couldn't see each other. But I couldn't think of anything to talk about, even to Peter. So, after about twenty minutes, we could hear a voice in the distance, saying 'Hallo. You're stuck!' Which we already knew of course. Then, a man in the lift with us started whimpering and I was worried that he was going to panic. I tried to find the control panel in case there was an emergency light but, I couldn't find it. Instead, I poked him in the eye by mistake. Well, at least he stopped whimpering.
FRANK	So how long were you stuck?
ROSALIND	Well, we then heard the voice again. 'Don't worry, we'll get you out of there very quickly.' It was very uncomfortable being with those people we didn't know. I was very nervous that I'd panic. Somebody made a joke, and I couldn't control my temper so told him to shut up. Fortunately, I was able to keep calm, but I couldn't express my feelings to the other people in the lift. In fact, it was at least forty-five minutes later before the lift started again. Then, as soon as we got out, we felt embarrassed. It's difficult to look at people after you've been in that kind of situation. Fortunately, I was able to laugh at myself afterwards, but, I didn't enjoy it much while it was happening.
FRANK	Well, I'm glad it all ended happily.
ROSALIND	Oh, it didn't end there!

Lesson 20 **Listening and writing, activity 4**

Part 2

FRANK	Well, I'm glad it all ended happily.
ROSALIND	Oh, it didn't end there! When we got back to the car, there was a car park attendant writing a parking ticket - £35 excess for being late back to the car.
FRANK	I don't believe it!
ROSALIND	I couldn't believe it either. I said to him, 'but we were stuck in your lift for an hour! We couldn't get back to the car in time.' He looked at me and said, 'I'm sorry, madam but you bought a two-hour ticket and your car has been here for three hours.' And then he said the one thing that *really* annoys me, he said, 'I'm only doing my job.'

FRANK	Oh dear, what happened?
ROSALIND	Well, I couldn't persuade him to take the ticket off the windscreen, so Peter started to argue with the man. I couldn't join in - I was speechless. They argued and argued, but the man said he couldn't take back the excess ticket now. So at that point, I got very angry, I couldn't keep my temper. What's more, I didn't have any money on me. And then Peter said he had plenty of money and was able to pay the fine. He actually got out the money from his wallet to pay the man. Well, I went mad, not just at the man but at Peter, as Well. I couldn't control myself. Usually, I can't tell people what I really think, because I can usually see both sides of the argument…
MAN	Very fair-minded!
ROSALIND	I shouted, I screamed, I stamped my feet. It only lasted about a minute, but it had the right effect. By now, there was a crowd around the car and suddenly I was able to take the ticket from the windscreen and I tore it up. The man was able to disappear into the crowd. We kept saying sorry to each other, sorry, sorry, sorry. But I wasn't sorry at all. I felt great. Just for once I lost my temper … and it worked. I feel great!

Lesson 21 **Vocabulary and listening, activity 4**

AMANDA	Oh, I wanted to watch Casablanca on the TV last night but I didn't get in time.
JOHN	Oh, you didn't get to see it.
AMANDA	No, did you see it?
JOHN	Oh, yes, it's a brilliant film, it's a wonderful film. Yeah, I watched it again, I watch it any time it comes on because it's a classic film, I never miss it.
AMANDA	Yes.
JOHN	Er … yeah, you should have seen it. I mean, the acting is fantastic …
AMANDA	Is it!
JOHN	… you've got Humphrey Bogart playing Rick who runs this bar in Casablanca on the north coast of Africa and Morocco. And, er … Ingrid Bergman comes in to the story, Clark Raines, Paul Heinrich they're all wonderful.
AMANDA	Hmm, and what about the plot? The plot's supposed to be very exciting, isn't it?
JOHN	Oh, it's an outstanding plot, really. It's er … got everything you need, it's a very emotional love story, Ingrid Bergman's married to this chap, and they come in, and she sees Humphrey Bogart, who was an old lover, and it starts the whole thing all over again, and, it moves, you know, in the midst of this strange setting, very gripping stuff.
AMANDA	Yes, and it's supposed to have a very powerful ending as well, isn't it?
JOHN	It does, it has a very memorable ending, indeed. It's an ending the image of which has been used in many commercials, it's that popular.
AMANDA	Yeah. So what would you say is your favourite scene in the film?
JOHN	Oh, I think the ending without question. It is … it's so moving, and Bogart is, er … and Bergman saying their farewell at this … at the airport, it's an image you never forget, it's, er … a love story that is … archetypal, it's outstanding. No question, it's something that you should certainly see.
AMANDA	I really must get to see it, yeah.

Lesson 23 **Listening**

MAN	Love is a wonderful romantic thing. Now do you think that Carol Ann Duffy would agree with that?
WOMAN	Hmm, not sure. I'm actually not sure whether she would or not from this poem. What do you think?
MAN	I think not.
WOMAN	Hmm.
MAN	No, I think not. I think that is the sense in which the poem expresses love as a rather dangerous and rather painful thing, a kiss is, shall we say … fierce.
WOMAN	And also, what is interesting there, is when she talks about giving the onion, its *fierce kiss*, it stays on your lips.
MAN	That's right.
WOMAN	But the onion is not what we would think of as the traditional symbol of love. So, um …
MAN	No, I think it's going for the reality of love, rather than the wonderfulness, the romantic …
WOMAN	I think so …
MAN	… the romantic, which is what the question was, wasn't it? The wedding ring is lethal which is a bit dangerous.
WOMAN	Yes, and she's still using her onion there, *shrinking its platinum loops* as she says. But, um … also …

MAN And, and, *its scent will cling to your fingers, cling to your knife,* which is again something that can … can … hurt you.

WOMAN Absolutely.

MAN She also says, *I am trying to be truthful.* She is not trying to be rom … as … maybe the opposite, she is not trying to be romantic, not trying to be loving, I am trying to be truthful. I am trying to express my feelings.

WOMAN Yes, and with that as well she rejects those sort of traditional ideas of Valentine's day gifts.

MAN Yes …

WOMAN … cutey cards or kissograms. Um … and I feel, in this, that, um … her gifts are symbols that are much more valuable than the cheap, um … little symbols such as these things, you know.

MAN But ultimately it's a … it's a cynical and rather sad view.

WOMAN She is cynical, she is cynical. But I think it's interesting, you know, that throughout this onion, she expresses her love as something deeper than this sort of cliched romance and red rose view of it.

MAN Sure sure. Um … love will make you cry. Do you think the poet would agree with that?

WOMAN I do, certainly do.

MAN Well, I agree. The onion … the onion … is going to *blind you with tears.*

WOMAN Yes, the onion again is so symbolic, onions make you cry, which would imply that her lover is, the onion, you know, she gives him an onion but in turn the two of them together will make her cry.

MAN Yes, yes.

WOMAN Very strong symbolism there.

MAN And the grief, that's a word she talks about, a word you use usually … describes the pain that you feel when someone's died or …

WOMAN Yes, it's a very strong word, *grief,* isn't it in this poem? And, and she says, um … as well, *onions would blind you with tears.*

MAN Yes, well, she definitely, um … she definitely thinks that love will make you cry.

WOMAN For sure, yes, yes.

MAN Even if a bit cynical. And love lasts forever. What do you think, do you think that the poem says that?

WOMAN Um, I think she would agree with that, although, in a sense I am not sure that she would see it as an on-going happy time. Do you know, what I mean?

MAN Well, I mean, I think the only clue we've got to that is … is where she says, um … *its fierce kiss will stay on your lips possessive and faithful as we are, for as long as we are.*

WOMAN Yes, absolutely, yes.

MAN Now, does that mean for as long as we live, basically till death us do part, or does it mean, for as long as we are in love.

WOMAN I think it could be interpreted in either way, actually, but what I think is important here, what I feel is, that she is saying how important it is to be faithful. So …

MAN Being expressed with an onion.

WOMAN Yes, yes. But, but the onion lasts forever because even if you cut it, you still have the rings, which actually, will maybe, shrivel to some smaller size, but you still *have* the ring.

MAN Well, it's organic they say. It will eventually … it will eventually, um … decay.

WOMAN Mmm, but I do feel that she's saying, um … that the scent is going to be there for an extremely long time.

MAN Yup.

Lesson 24   Speaking and listening, activity 2

FRANCIS So tell me, Shelley, I'm curious about the media in America. Do people read mostly regional or national newspapers over there?

SHELLEY I think that really most people read regional magazines or … or newspapers, um … although the national news and the international news is reported in the regional newspapers through the wire service.

FRANCIS Right, right. Right, I see. And … and as far as reading habits goes as well, do most people read a newspaper every day or … ?

SHELLEY You know, I think a large percentage of people do, but it may just be a local newspaper that they're reading, maybe from their town or the town … a larger town nearby.

FRANCIS Right, right. Well, as far as, er … the broader media is concerned, I mean, how about television, I mean for example, I know that there are a lot of television channels, but how many television channels are there?

SHELLEY Well, that really depends on where you live. In cities there are dozens and dozens of channels that you can tune into.

FRANCIS Right.

SHELLEY Yeah, but in the rural areas, er … sometimes you'll only get the three major national channels and perhaps the public broadcasting.

FRANCIS Oh, really, aha. And what sort of, what about the programmes themselves. What are the most popular types of programmes?

SHELLEY Well, I think probably soap operas are always the most popular, but …

FRANCIS Just like here.

SHELLEY Yeah, but a few years ago, er … game shows were very popular and in the last couple of years chat shows or talk shows have become very popular and you can watch a talk show, talks shows that will go on from the morning all the way straight through to late at night.

FRANCIS You're kidding, really.

SHELLEY No, no kidding.

FRANCIS My goodness me. We've got a little way to go to catch up with you there. But what about, um … radio, for example. I mean … is … would you say radio is more or less popular than television?

SHELLEY Oh, I think television's the most popular. But, er … it's different, I think than in Great Britain, radio stations in America are mostly just music stations, there's a wide variety of different kinds of music and there is a national broadcasting, national public radio it's called, that has news and plays. But most radio stations are just music.

FRANCIS Right, right. I mean are there … are there government restrictions, what about … I mean … you said about public broadcasting but I don't know if that's government. But … what about … do you think that the government in any way controls the media?

SHELLEY Gee, I … I really hope that they don't. We're told that they don't, but, er … you never know, it depends on the information that's being given by the government to the newsmen. We have a history of free press.

Lesson 26   Reading and listening, activity 2

Q So what is Scrapple?

CHEF Scrapple is a dish from New England and the basic ingredient is pork. It was a dish that was made to use up all those bits of pork which were left over. It's cooked with onions and cornmeal, and salt and pepper, of course. It's hearty farmer's meal.

Q This is a wonderful name - Grand Central Oyster Stew.

CHEF Well, that's a classic New York dish, made with oysters, butter, milk and cream. Very tasty. It was first made by the chef at the Grand Central Station in New York.

Q And Jambalaya?

CHEF From the Deep South. Hot and spicy, a mixture of pork, sausage, ham, prawns, tomato purée, onion, garlic and chilli and served with rice.

Q And what is Tacos?

CHEF What **are** Tacos? Tacos are tortillas filled with minced beef and pork and with lots of chilli and other spices, such as cumin.

Q What is Tex Mex?

CHEF It's the term for the cooking in that region. Texas and Mexico. Don't forget that Texas was part of Mexico once.

Q And what about Cioppino?

CHEF Ah … that's a dish from the West Coast with fish, mussels, crab meat, prawns, tomatoes, wine and herbs. Delicious.

Lesson 26   Reading and listening, activity 4

CHEF And the last dish is hashed browns, from New York.

Q And how do you make that?

CHEF Well, it's basically potatoes. You start by heating the butter and the oil in a frying pan. Oh, make sure you use a heavy frying pan, or it will stick. While you're waiting for the butter and oil, mix the potatoes and the onions in a bowl. Then, when the butter and oil is hot, you spread the mixture over the bottom of the pan and press it flat. Never heat the oil too much as it'll start smoking and that can be dangerous. Always spread the mixture evenly so it cooks all the way through, and make sure you press it down into the pan. Don't forget to add pepper and salt on the top and cook for about eight minutes. Check that it is brown underneath, then turn it over. Then, add the cream and cook the other side for about ten minutes. And there you go. Oh, I almost forgot. Always serve it immediately, as it'll spoil if you have to keep it warm. It's a great dish, a Yankee classic.

Q What do you eat it with? More to the point, when do you eat it?

CHEF It's a breakfast dish, and you eat it with eggs and ham, and toast and breakfast things like that.

Lesson 27 **Listening and writing, activity 2**

SPEAKER 1 I think that's Budapest, um … to me it's the most cultured place on earth. You can go to the opera very cheaply, you can go to the Turkish baths, and you can go to beautiful cafes, and it's as if time stands still, and it's still like a hundred years ago.

SPEAKER 2 Er … well, when I was in New Zealand I did a bungee jump, and, er … it's just like jumping off a bridge into a river except, you get held back by this long piece of elastic. And I screamed the entire time, it was terrifying.

SPEAKER 3 Well, without doubt, when I was on holiday in Italy, I was lucky enough to be taken to the racing circuit at Monza. And, um … I was in a proper racing car with an instructor, going round the course at great speed. It was thrilling. It's something I've always wanted to do and would love to do again. Absolutely thrilling, the speed: exhilarating.

Lesson 30 **Reading and listening, activity 3**

1 Well, the problem is that the pool doesn't have the visual opportunities that a running track has, and to make swimming more exciting, I'd make them use a circular pool and make them try to pass each other, so that it's more like a running race. It'd be easier to see that they are competing against each other.

2 Well, football is a very fast game now, much faster than it used to be, and the referee cannot be expected to see what is happening all the time, and sometimes they make mistakes - after all, they're only human. So, I would let the referee look at a video recording before he decides to send a player off. It's very hard to make up your mind if you haven't seen exactly what happened.

3 Well, there aren't many games in which you can get scores of over a hundred on occasion. Someone scores every two or three minutes. I wouldn't let people over 2 metres play basketball. At the moment, it's very easy for the players to score, so it isn't very exciting to watch. If the players were shorter or the basket higher, it would make them work harder.

4 Boxing is very dangerous and anything to make it safer can't be bad. Why don't they make them fight in a round ring, not in the traditional square? It's quite usual for a fighter to get trapped in the corner at the moment. If the ring was round, it would be safer, I reckon.

5 For the male players, tennis is just a game of strength now, there's little skill involved. To bring back an element of skill, I … I wouldn't let the players have a second serve. If they serve unsuccessfully once, the opponent should get the point. It's likely to need more accuracy and less power.

6 Skiing can get very boring for the spectators. I'd let them go down the slope all at the same time, racing against each other. It's boring to watch them one after the other, racing against the clock and not against each other.

7 The trouble with football is that no one scores very much. A match with three or four goals is quite rare these days. And it's the goals which makes football exciting, in my opinion. But if the goal posts were wider apart it would let players score more. At the moment they're quite narrow, and it's difficult for players to score very often in a match.

8 Tennis is too fast these days. I think they should make them play with heavier rackets. It's essential to slow the game down. We want to see how well they can play, not how strong they are.

Lesson 31 **Vocabulary and listening, activity 2**

SPEAKER 1 I always take a spare key because I'm incredibly forgetful and I very often leave my keys at home, and pull the door closed behind me, with the keys still on the kitchen table. So, if I take a spare key, I don't have to disturb my neighbour, who also keeps one for me.

SPEAKER 2 If there's a long wait, I like to have something to read. So I always take a book in my bag in case I have to wait for a bus or something. Even in the cinema, you can wait ages for the film to start.

SPEAKER 3 Well, it sounds a luxury because I certainly don't need one for my job, but the mobile phone makes me feel much happier in case I have an accident or in case I'm delayed. I've got one of these small ones which fit into my handbag and hardly takes up any room at all.

SPEAKER 4 I'd be completely lost without my diary. It contains all my addresses as well. If I'm out all day on business and can't remember what my next appointment is, I check in my diary.

SPEAKER 5 I never go anywhere without a bar of chocolate. I always have one in my bag in case I get hungry. The trouble is, of course, is that just the idea of having the chocolate in my bag makes me feel hungry, so then I eat it, and panic that I haven't got any more, so I go out and buy another bar, and so it goes on.

SPEAKER 6 I never leave without a pile of business cards. They're useful if you have to leave someone a note, or in case you need to make sure that the person you've been seeing remembers who you are.

SPEAKER 7 I always take a notebook and a pen of some kind, usually a ballpoint because I seem to lose my good pens. I like to write short stories and I need to have something to write with in case I want to make notes.

SPEAKER 8 I always have a penknife with me. Actually, it's a special penknife, it was my father's and its got a lovely handle. But it's useful if I need to cut something like fruit or cheese, or open an envelope.

Lesson 32 **Reading and listening, activity 2**

MR CLARKE Hello, Clarke's Garage.

MR FISH Yes, hello, it's Hector Fish here. I'm just ringing to find out what's happening over this business with my car. Have the parts arrived yet?

MR CLARKE Ah, Mr Fish. Er … yes. Well, let me just have a look at the file.

MR FISH You don't need to look at the file. I wrote to you on the 16th of July, and you still haven't replied.

MR CLARKE I'm not sure if we've received, um … ah, yes, here it is. Well, we're still waiting for the parts, and they're due to arrive in, er … let me see now, er … yes, in three days' time. We'll let you know as soon as we get them, and we'll arrange a time for you to bring the car in to us.

MR FISH I really don't see why I should have to bring it to you.

MR CLARKE No, you're quite right, er … but we are still very busy and if you can bring the car yourself, you'll be helping us very much. I'm very sorry this has all happened, we're just doing our best to get it sorted out. So, I'll be in touch with you as soon as we receive the parts. Is that all right, Mr Fish, Mr Fish?

Lesson 33 **Listening and writing, activity 2**

If your new leather shoes feel a bit tight, pour a small amount of boiling water into each shoe in turn. Leave them for no more than six seconds, pour the water away and put on the shoes. The leather will have softened sufficiently for your feet to stretch them so that you no longer feel the pinch.

To stop hiccups, block your nose and ears with your fingers and sip water at the same time. Someone else has to hold the glass, of course, so if it doesn't stop the hiccups at least everyone has a good laugh.

If you have made a mark on a piece of wooden furniture, for example, a watermark, you can remove it by mixing some cigarette ash with some olive oil until it forms a paste. Spread this over the mark generously and leave it overnight to remove it. Next day the mark should have disappeared.

To remove chewing gum from an item of clothing, place it in the freezer overnight. The following day, the chewing gum can be easily scraped off. The same principle applies to gum on carpet. Rub it with an ice cube until the gum is solid and then scrape it off.

Lesson 34 **Listening and speaking, activity 3**

MAX So, when do you think the best time to go is?

SUSIE Well, it's always the matter of weather. I mean, if we visit Nepal in April, we'll meet fewer tourists because the main tourist season is in the summer. But if we go in the summer, say, in August, we'll have better weather.

MAX Well, I'd like to avoid the tourists, but if we go there out of season, we'll probably have bad weather.

SUSIE That's right.

MAX And how do we get there?

SUSIE I think the best is to fly. It really is a long way from here. And if we fly, we'll get there much quicker.

MAX But if we drive overland, it'll be very interesting and we'll be able to travel around more easily when we're there.

SUSIE Yes, but it's going to take weeks to get there by car, and I haven't got the time.

MAX Do you want to stay in hotels?

SUSIE If we stay in hotels, we'll spend a lot of money. No, I think it'll be better to go camping. If we take a tent, we'll save a lot of money.

MAX What about food? Shall we take our own food?

SUSIE If we do that, we'll have a lot to carry. Oh, there's something else which is quite important. If we pay for the holiday here in Britain, we can use British money, and we don't have to take so much with us. But, if we pay for it there, it might be cheaper.

MAX Well, it will be cheaper if we take a guide book. That way we'll avoid hiring a local guide.

SUSIE Yes, and we can read up about the place before we go there.

Progress check 31–35 **Speaking and listening, activity 3**

GERTRUDE Er … now, I was told there are certain ways to behave, when I go to Britain. And I would just like to check a few of those statements with you, because I'm not quite sure, um … if they … if they really apply.

JOHN Okay.

GERTRUDE And I would like your advice on those. For instance, if I enter a railway carriage, um … am I supposed to shake hands with everyone?

JOHN No, no, the English tend to keep themselves to themselves … No, they would think you were mad if you went and shook hands with everyone.

GERTRUDE Really?

JOHN Yes.

GERTRUDE So, it isn't a good idea?

JOHN Not a good idea.

GERTRUDE And, um … I was also told that, if I go to a greengrocer's I can pick up all the fruit and choose the best.

JOHN I don't think they, er … would like you to do that because if everybody did that then the fruit would get very squishy, and, er … nobody would want to buy it. So, no, not on that one, no.

GERTRUDE No. And, um … if I go to a pub I was told I have to go straight to the bar, order my drinks and I have to pay for them immediately. Is that true?

JOHN Yep, that's right. If you were in a cafe or a restaurant then you'd pay it … pay for the drinks later, but in a pub you pay for them, straight away.

GERTRUDE So … and if I don't do that?

JOHN Um … they would make you.

GERTRUDE Oh, and staying in the pub, um … I was told that the best way to actually make friends and get to talk to people is to say can I buy everyone a drink? Is that so?

JOHN Well, you'd be very popular, but you might also get very poor very quickly if it was a large pub. But certainly, no you would be very popular.

GERTRUDE Right. I'll have to try that.

JOHN Come to my pub.

GERTRUDE I will. And, um … I was also told, and this sounds quite strange to me, that if I need help in a shop, if I want the attention of the shopkeeper or sales assistant I have to clap my hands.

JOHN Oh no, no, you would um … be very unpopular if you did that. It smacks of, um … being a mistress and a servant relationship.

GERTRUDE Is that an old custom, would that have been the case a hundred years ago?

JOHN Maybe, maybe, yes. But not now.

GERTRUDE Right. And, um … somebody told me that taxi drivers, um … and shop assistants can call a girl 'love' or 'darling'.

JOHN Oh sure, yes, that's very common. 'Hello love', 'all right darling'.

GERTRUDE Does it mean anything?

JOHN It means, they're being friendly, it means they're being friendly and nice, yeah. It's complimentary.

GERTRUDE All right that's very nice. And, um … now that surprised me the most. I was told there was a brilliant campsite for caravans and tents actually in the gardens of Buckingham Palace.

JOHN Buckingham Palace!

GERTRUDE Yes, is that true?

JOHN No, no, you were being wound up. You were being teased. That was a joke. No, no, the grounds of Buckingham Palace are for the Queen and the Royal Family, only.

GERTRUDE Oh, what a shame!

JOHN I'm afraid so.

GERTRUDE Hmm. And, um … I was also, I got this information that people who are … who get stuck in a traffic jam usually sound their horn.

JOHN Some people do but, um … not … the English tend to keep off the horn. In Italy … you sound your horn in Italy, you sound your horn off the Peripherique in Paris. But you get very unpopular if you sit in a traffic jam and sound your horn. No.

GERTRUDE It's not polite.

JOHN No.

GERTRUDE Right. I won't try that. And, um … and also is it true that it is really totally normal and acceptable to actually read someone else's newspaper over their shoulder.

JOHN Well, I think it rather depends what sex you are. I think if I was reading a newspaper and I had a man reading my newspaper over my shoulder, I wouldn't like it, but if it was a gorgeous looking girl, I kind of wouldn't mind.

GERTRUDE Oh, right. So, there is a difference!

JOHN Yes, I'm not too sure about that one. But I think probably in general, no. But there are exceptions.

GERTRUDE It's better not, but it is not a disaster if you do.

JOHN Do it through dark glasses, then they can't see your eyes.

GERTRUDE All right. And, um … when you say goodbye to people, do they really always say 'Have a nice day'?

JOHN Americans do. Um … it's becoming a little bit more of a habit in England because of the American influence. But I don't like it, lots of people don't like it. But it's a very American thing – 'Have a nice day now'.

GERTRUDE Yes, right. So, it's not a really British thing to do.

JOHN Not really British, no.

GERTRUDE Right, thank you.

JOHN Okay.

Lesson 36 **Listening and writing, activity 2**

MAN It's a bit difficult that one, isn't it?

WOMAN Yeah, um … I suppose what I would do is imagine how I'd feel if I was in his situation.

MAN Hmm …

WOMAN And think, well, I would really want a friend if I was feeling upset or depressed and I'd called round to see somebody for a shoulder to cry on. I would want them to be there for me, so I suppose I'd just think, well, okay, so I'm feeling tired, and it's the beginning of the week, and it's ten o'clock but um …

MAN But would you be sensitive to the other person if you were in … if you were in a state and went round for comfort would you see that … whether or not they, er … were ready to talk or whether you could see in their eyes that they wanted to go to bed and didn't want to talk.

WOMAN Um … I don't know really, I would hope that if it was a close friend, that, that … I … I wouldn't have to worry about that.

MAN They'd drop everything.

WOMAN Yeah, yeah, because I'd expect it of my friends, so I think I would do it for them.

MAN Do as you would be done by.

WOMAN Yes, I think so.

MAN Yeah, I agree.

Lesson 37 **Listening, activity 1**

Q So why did you choose Hong Kong then for your holiday?

JILL Well, my brother lives there and, um … I wanted to go out there because we are twins, in fact, and we're were celebrating our fortieth birthday.

Q Oh, right. When did you go?

JILL In the summer. It was so hot, it was unbelievably hot.

Q Yeah, brilliant. Whereabouts did you stay when you were there?

JILL Well, I stayed actually with my brother on Cheung Cheu Island, um … which was very cheap, it made it a very cheap stay. But, um … I actually wish I'd spent some time in Hong Kong Island or in fact in Kowloon.

Q What … what sort of things did you do when you were there?

JILL Oh, sightseeing, um … it was brilliant, we went up the peak, Victoria Peak it's called as you probably know, um … we visited Aberdeen like everybody did and um … it was just wonderful, we saw, I think everything that a tourist should see.

Q Brilliant. Was it boiling? I mean were you in shorts and … ?

JILL All the time, all the time, and in fact I wore as little as possible and I … I wore flip-flops on my feet, and in fact, um … made the mistake of going into a restaurant in them which is a big *no-no*.

Q Oh really!

JILL Yes, in fact, you know, one is supposed to wear them in private but certainly not out in public.

Q So, and … was it good food in the restaurant?

JILL Terrific, terrific! What I really loved was dim sum, um … I tried it one day, when I was alone, and then, in fact, felt more adventurous and ate it in restaurants, but I really overate, it's very easy to actually eat too much because, things are in little portions, so trying lots of little things you could soon consume too much.

Q Did you buy anything else when you were there? Did you do any shopping?

JILL I certainly did. I spent far too much money. Silk is just beautiful and relatively cheap. I bought silk dressing gowns for my children, a camera for my husband as instructed. But I spent far too much money.

Q Well, you've sold me, I'm going to go now.

JILL Do go.

Lesson 39 **Vocabulary and listening, activity 2**

FRANCES A kindergarten. Well, in Britain that's for children between the ages of three and five and is also called nursery school, actually.

JOHN In America, kindergarten is the first place that a child can go to school and it's voluntary and it's for a child of aged five not younger than that and, er … you just attend for that one year before you start the first grade.

FRANCES Oh, right. Well, it's funny you should say the word 'grade', because in Britain grade means a score or a mark, like if you handed in some work you'd get a grade A, B or C for it and so forth.

JOHN Well, that's also true in America except that grades also correspond to the level you are in school, for the first, in your primary school years, first, second, third, fourth and up to sixth, seventh and eighth grade. It's the year you are in.

FRANCES Oh, right. Yeah, a primary school is for children from the age of five up until nine or eleven sometimes. And it's often divided into infants and junior school.

JOHN Hmm. Well, our equivalent would be grade school or elementary school and as I said you start then at first grade after kindergarten, you go to fifth or sixth grade before you start the next level.

FRANCES Oh, right. Um, well I mentioned junior school, that's often called middle school sometimes in some counties and that's for the age of nine up until thirteen.

JOHN Right, I suppose our equivalent would be, somewhere around about junior high school, which you start sometimes depending but usually about … seventh or, eighth grade.

FRANCES Oh, right. Um … well, the next one after that, would be secondary school in Britain and you'd start there from the age of eleven or thirteen and go up until the age of sixteen.

JOHN Right, well the equivalent for us in … at that level, would be high school, although I think we go a year more, or maybe two years more seventeen, roughly, eighteen years to finish high school. That's through to twelfth grade.

FRANCES Right, yeah. You can also get a comprehensive school which is the state secondary school.

JOHN Hmm. State school in America is just a public school, you know a state secondary school or primary school, a public school open to anyone in the public.

FRANCES Yeah, it's funny you should mention that because in Britain a public school is actually a private school, um … and it's for children, er …whose parents pay fees for them to stay there so anyone who can afford it can go.

JOHN Right, well we don't … there are very few private schools in that way other than say, parochial Catholic schools and military academies, and stuff like that. But … no, almost all the schools in America are public schools.

FRANCES Oh, right. Of course the other one to mention is boarding school which is a school where pupils can stay overnight, and can go home at the weekend or sometimes at the end of term. And some children go to boarding school, you know, as early as eight years old.

JOHN Oh wow!

FRANCES And, of course, the school leaving age generally in Britain is sixteen.

JOHN Yes, I think that's the same. I think you can leave at, er … sixteen, you don't have to finish high school but you have to do up to, I think, about the second year of high school, aged sixteen.

FRANCES Yeah.

Lesson 39 **Vocabulary and listening, activity 4**

Q Well, I suppose the schooling system in America and England is not quite the same way I would imagine. For example, John, what age do you start school?

JOHN In America, most people start school at … age five. They go to kindergarten, and it's a voluntary thing but most people do do that.

Q At five, yes.

FRANCES I actually went to state primary school and I started just after my fifth birthday, um … you know most people start at five and then they can leave when they're sixteen. I actually stayed on until I was eighteen.

Q So, basically it's more or less the same starting age. But, er … when do you finish in the States?

JOHN Well, we finish, er … you can go to high school, most people finish high school, but you can actually leave school legally at the end of the tenth grade which is about sixteen … when you're sixteen.

Q So, more or less actually the same thing. And, er … as far as examinations are concerned, I mean, is it, um … I think it's very different, isn't it?

FRANCES Well, I don't know. I took exams at the age of seven, then ten, then fourteen and then sixteen.

Q And what about in America?

JOHN Well, exams are not quite the same, they don't have the streaming exams in America that they do in Britain. So, we have exams right throughout our educational career. You know, from the time you're in the first grade you have tests and things all the way up through, but not with the same kind of pressure on you. As they have in Britain.

Q I see, yeah, it's easier, isn't it, in America? Yeah …

JOHN In that way, yeah.

Q And … and in England, I mean, England is famous for, er … punishments, isn't it? I mean … I don't know if you do it still, but is it still going on?

FRANCES I didn't really have any horrible punishments at school, I think sometimes at some schools it does go on. Um … we had things like detention and lines, you had to write lines out for a hundred times or something like that.

Q But no beatings or things like that?

FRANCES No, no, no not for me, anyway.

Q And in America?

JOHN I think when I was very young, because that was some years ago they may have had light corporal punishment, but I think that all of that has been abolished now, they don't have it. They are not allowed to.

Q For the best, yeah. And homework, I believe that homework in England is quite like, er … a serious thing, no?

FRANCES Yes, I think the older you got the more homework you got. Um. … I think probably I had about half an hour's homework a day on average.

Q That's not too bad.

FRANCES Not too bad, no.

JOHN That's very good I think I probably had more, I think on average around about an hour a day.

Q Oh, really, I thought that America was actually easier on, er … on homework?

JOHN Well, it depends, but you start getting into high school and you have certainly an hour a night.

Q Oh, really, OK, and, er … I mean, I think I understand that in America the classroom participation is a very, very important feature, isn't it?

Q Oh yes, that's very true. You … you get to know, there's a more informal kind of, er … relationship, I think, between the school kids and the teachers in America than perhaps there is here, certainly as there was before, maybe that's changing now.

FRANCES Right.

JOHN And I think that contributes a lot because you feel more free to ask questions.

Q That's right, and that tends to … that tends to give you assertiveness, doesn't it?

JOHN Yes, it does help.

FRANCES Well, I mean … we had very little classroom participation, um … we had sort of group work with a teacher, you know you have question and answer sessions.

Q But more formal than …

FRANCES Yeah, a lot more formal, actually.

Q And, I mean, I don't know what the situation in America and England, I mean, do you have to pay for school, universities, and things like that?

JOHN Oh yes, yeah, but for example, me I went to university, at a state college, a state university and if you … if you … were a resident of the state then really you don't have to pay as much as someone, say coming from outside the state would. It still costs a fair amount. It still costs a fair amount.

FRANCES Yeah, I went to university for three years, and I was very lucky because I had my fees paid by the local education authority, I think that's becoming less and less common these days.

JOHN My parents had to meet my tuition charges, because your fees are fairly steep but as I say they are not as bad as private universities.

Q And what about getting in there?

JOHN Into university?

Q Yeah.

JOHN Well, in America, it depends on what your grade point average is coming out of high school, it depends on certain examinations that you take, test scores outside of the school system. What they call the S.A.T. scores and various achievement tests.

FRANCES What happens, you choose a university that you wanted to go to and then that university would make you an offer or may not make you an offer, um … I was very lucky and I went to the university of my choice, but they do ask for certain grades. Um … I think … I think it was like three B's when I went. But I think it's getting higher and higher now, and harder and harder.

Q So more selective.

FRANCES Yes, yes.

Q Are there, is there anything that you regret about your education? I mean, your training, things like that?

FRANCES Um … well. I really regret not learning how to cook. I'm a terrible cook. And also I regret not learning how to type, because I think typing is a very useful skill to have.

Q Indeed, yes.

FRANCES The other thing is, that I think, you know, I went straight to university from school, I didn't have any time off, and, er … I think I should have taken a year off maybe and gone travelling. I feel I went to university too young.

Q Hmm. And what about you, John?

JOHN Well, I regret, I suppose, that university cost my parents so much, as they paid my tuition fees it was quite hard for them in those years. It would have been nice to have a scholarship to help out. Also, I suppose, I didn't work quite as hard as I should have done, I wish I'd worked harder. But, er … it was a grand time to be, and I have to be honest and say, it was the happiest time I can remember, and I wouldn't mind still being a student.

Q All right. Thank you very much, both of you.

Lesson 40 **Reading and listening, activity 2**

Part 1

He first noticed the new man in the neighbourhood on a Tuesday evening, on his way from the station. The man was tall and thin, with a look about him that told Ray Bankcroft he was English. It wasn't anything Ray could put his finger on, the fellow just looked English. That was all there was to their first encounter, and the second meeting passed just as casually, Friday evening at the station. The fellow was living around Pelham some place, maybe in that new apartment house in the next block.

But it was the following week, that Ray began to notice him everywhere. The tall Englishman rode down to New York with Ray on the 8:09 train, and he was eating a few tables away at Howard Johnson's one noon. But that was the way things were in New York, Ray told himself, where you sometimes ran into the same person every day for a week.

It was on the weekend, when Ray and his wife travelled up to Stamford for a picnic that he became convinced the Englishman was following him. For there, fifty miles from home, the tall stranger came striding across the rolling hills, pausing now and then to take in the beauty of the place.

'Linda,' Ray remarked to his wife, 'there's that fellow again!'

'What fellow, Ray?'

'That Englishman from our neighbourhood. The one I was telling you I see everywhere.'

'Oh, is that him?' Linda Bankcroft frowned through the tinted lenses of her sunglasses. 'I don't remember ever seeing him before.'

'Well, he must be living in that new apartment in the next block. I'd like to know what he's doing up here, though. Do you think he could be following me?'

'Oh, Ray, don't be silly,' Linda laughed. 'Why would anyone want to follow you? And to a picnic?'

'I don't know, but it's certainly odd the way he keeps turning up … '

It certainly was odd

Lesson 40 **Reading and listening, activity 3**

Part 2

And as the summer passed into September, it grew odder still. Once, twice, three times a week the mysterious Englishman appeared, always walking, always seemingly oblivious of his surroundings.

Finally, one night on Ray Bankcroft's way home, it suddenly grew to be too much for him.

He walked up to the man and asked, 'Are you following me?'

The Englishman looked down his nose with a puzzled frown. 'I beg your pardon?'

'Are you following me?' Ray repeated. 'I see you everywhere.'

'My dear chap, really, you must be mistaken.'

'I'm not mistaken. Stop following me!'

But the Englishman only shook his head sadly and walked away. And Ray stood and watched him until he was out of sight.

'Linda, I saw him again today!'

'Who, dear?'

'That Englishman! He was in the elevator in my building.'

'Are you sure it was the same man?'

'Of course I'm sure! He's everywhere, I tell you! I see him every day now, on the street, on the train, at lunch, and now even in the elevator! It's driving me crazy. I'm certain he's following me. But why?'

'Have you spoken to him?'

'I've spoken to him, cursed at him, threatened him. But it doesn't do any good. He just looks puzzled and walks away. And then the next day there he is again.'

'Maybe you should call the police. But I suppose he hasn't really done anything.'

'That's the trouble, Linda. He hasn't done a single thing. It's just that he's always around. The thing is driving me crazy.'

'What - what are you going to do about it?'

'I'll tell you what I'm going to do! The next time I see him, I'm going to grab him and beat the truth out of him. I'll get to the bottom of this … '

Lesson 40 **Reading and listening, activity 7**

Part 4

Some time later, the tall Englishman peered through a cloud of blue cigarette smoke at the graceful figure of Linda Bankcroft and said, 'As I remarked at the beginning of all this, my darling, a proper murder is the ultimate game of skill … '

Macmillan Education
Between Towns Road, Oxford OX4 3PP
A division of Macmillan Publishers Limited
Companies and representatives throughout the world

ISBN-13: 978-0-435-24024-0
ISBN-10: 0-435-24024-2

Author's Acknowledgements
I am very grateful to all the people who have contributed towards *Reward*
Intermediate. My thanks are due to:
– All the teachers I have worked with on seminars around the world, and
  the various people who have influenced my work.
– James Richardson for producing the tapes, and the actors for their voices.
– The various schools who piloted the material, especially Michelle Zahran at
  Godmer House, Oxford School of English, Oxford; Jon Hird at the Lake
  School, Oxford; Philip Kerr at International House, London.
– The Lake School, Oxford and especially Sue Kay for allowing me to observe
  classes.
– The readers, all of whom wrote encouraging and constructive reports.
– Simon Stafford for his first class design.
– Jacqueline Watson for researching the photos so efficiently.
– Chris Hartley for his continual advice and encouragement.
– Angela Reckitt for taking over as editor so calmly and effectively.
– Catherine Smith for her thorough and considerate management of the
  project.
– and last, but by no means least, Jill, Jack, Alex and Grace.

Designed by Stafford & Stafford

Cover design by Stafford & Stafford
Cover illustration by Martin Sanders

Illustrations by:
Adrian Barclay (Beehive Illustration), pp3, 22, 23, 26, 37, 73;
Hardlines, pp27, 76;
David Mostyn, pp8/9, 12/13, 46, 96;
Martin Sanders, pp18/19, 24/25, 37, 40/41, 44, 48, 56/57, 58, 63, 64, 66, 67,
68/69, 78, 85, 90, 94;
Simon Stafford, p10.

Commissioned photography by:
Roger Charlesson pp20/21, 30(Batman), 54/55, 90;
Chris Honeywell pp8/9, 14, 34, 35, 74/75.

Acknowledgements
The authors and publishers would like to thank the following for their kind
permission to reproduce material in this book:
Rogers, Coleridge & White Ltd for an extract from *The Tropical Traveller*
© John Hatt, 1982; Reed Consumer Books for an extract from *It can't be true*
and *The world's greatest mysteries* © Octopus Books; *The Independent on
Sunday* for an extract from 'A Pilgrim's Package' by David Lodge and 'Nothing
beats a good rant' by Rosalind Twist; Anvil Press Poetry for the inclusion of
'Valentine', a poem from *Mean Time* by Carol Ann Duffy; *The Observer* for an
extract from 'Memories are made of this' by David Randall and 'Breaking the
rules' by Norman Harris; Premier Magazines for an extract from 'All dressed in
red' by Alexandra Davison and 'Wild and beautiful' by Brian Jackman, first
printed in British Airways *High Life* magazine; Hamish Hamilton Ltd for an
extract from *A Year in Provence* by Peter Mayle © Peter Mayle, 1989; Michael
Joseph Ltd for an extract from *Superhints* compiled by The Lady Wardington;
Mandarin Paperbacks (an imprint of Reed International Books) for an extract
from *The Good Tourist* by Katie Wood and Syd House; Random House UK
Ltd, for an extract from *Beyond Belief* by Ron Lyon and Jenny Paschall;
Randall Brink for the inclusion of an extract from 'Life Stories: Amelia Earhart;
the unsolved mystery', as published by *Marie Claire* magazine, June 1994;
Lonely Planet Publications for an extract from *Hong Kong - a travel survival kit*
by Robert Storey (May 1992).

Photographs by: Aquarius p30; Art Directors pp38/39; The Anthony Blake
Photo Library p62; Micheal Busselle p44(t); Mary Evans Picture Library
p82(detail); Sarita Sharma/Format p6; Joanie O'Brian/Format p42; Maggie
Murray/Format p43; Chrispin Hughes p28; The Hulton Deutsch Collection Ltd
p82; The Image Bank pp32, 86; Images Colour Library pp42, 92; The Kobal
Collection p31; National Film Archive pp50/51; Pictor International p92;
Science and Society Picture Library p16(t); Catherine Smith/Matthew
Sherrington pp80, 81; Spanish National Tourist Office p44(b); Simon Stafford
p76; Tony Stone Images pp32/33, 42, 43, 52, 70(a), 70(b), 88, 89; The
Telegraph Colour Library pp70(c), 98; Topham Picture Source p16(b); Eye
Ubiquitous p42; Jacqueline Watson p80(t); Zefa Picture Library p4.

The publishers would also like to thank Becky Jones, Daphne Levens and
Guy Lowe.

Printed and bound in Spain by Mateu Cromo
2009 2008 2007 2006
26 25 24 23 22 21 20